Mathematical Snippets

Exploring mathematical ideas in small bites

3.141592

6358979323384

theoni pappas

Wide World Publishing

DISCARD

Some topics in *Mathematical Snippets* were originally published in non-book form.

Wide World Publishing/Tetra
P.O. Box 476
San Carlos, CA 94070

website:
http://www.wideworldpublishing.com

1st Printing June 2008

ISBN 10: 1-884550-39-8
ISBN 13: 978-1-884550-39-3

To the mathematician in all of us.

It's there.

Do the math!

TABLE OF CONTENTS

ix   What's a mathematical snippet?
2   The disappearing columns of St. Peter's Square
4   Who really chooses whom-x, y or z?
5   Digital forensics
7   Boids & Birds
8   Getting a glimpse at the mathematical world of Salvador Dali
12   Math or magic?
13   The Alhambra & The American Institute of Mathematics
14   Einstein's cosmological constant blunder?
15   The big box for small fabrications
14   Who really chooses whom-X, Y or Z?
15   Bridges that go anywhere
16   Dancing with the numbers
17   Bridges that go anywhere
18   Networking is not schmoozing
21   Mathematics for math's sake
22   Zeno's "zany" paradoxes & more
24   A closer look Zeno's Dichotomy, Achilles & the tortoise, Stade paradoxes
25   Riemann hypothesis and more math surfaces in The Case of Lies
27   Polling & mathematics
28   Juggling mathematics
30   What's so special about random numbers?
32   The elusive black swan
34   Voting paradoxes
37   The 4-color math problem's quirky twist
38   The road to chess
39   Computers & random numbers
41   The Grand Canyon & randomness
42   The problem that determined Gauss' career
44   π's rock star status
45   A moment in the life of data searches
46   Geoengineering
47   The eyes that launched the hearts of countless readers
48   The Leonardo da Vinci Museum
50   To be or not to be Rembrandt
52   How bees and other insects fly
53   Universes of the past
55   Mathematics behind the perfect golf ball
56   Mathematics caps off the dome
59   Are plants counting?
62   Are you up to the second?
65   Whodunit?…Jackson Pollock & fractals
68   Math & your ring tones
70   Surfing with mathematics
71   Tackling Murphy's Law with math
72   The Sierpinski triangle finds its way into the Grand Egyptian Museum
76   The misplaced manuscripts fiasco
77   The inheritance problem
78   The Möbius strip continues to twist our fancy
81   Saccheri missed the boat
82   Points of deception
85   The sound of money
86   Witricity may cut the chord

87 How computers play chess
90 What powers mathematics?
92 Some things don't change- Euler's polyhedra theorem
93 Mathematics unravels quasicrystals
95 Nicomedes' conchoids goes green
96 Constants, constants everywhere
99 A mathematical oasis
100 4 deceptively simple blockbuster problems
101 The Turing test
102 The four forces of the universe
103 Nanotechnology & the bridge to outer space
104 2 is odd
106 Hacking into communication of bacteria
108 Branes, gravity and all that stuff
110 Mathematical knots may unleash computer power
112 Multiverse theory & fractals
114 Mathematics & patterns
116 Pozzo's trompe l'oeil
117 Hear no, see no, speak no new mathematics
118 Logic problems
119 Puzzlers
120 Famous mathematical goofs
122 Something from nothing?
123 The Fibonacci waltz
124 Quasi-periodic music and more
126 The mysterious Antikythera mechanism
130 M-Theory
134 Dissections-cutting up mathematics
135 The impossible formula & Abel's life
138 Nanotechnology-then and now
140 Beware of nano-hazards
142 Nanotechnology watch dog?
144 Balancing problems
145 Out, out damn nine
146 Global warming & the math behind it
150 Graphs-the picture of global warming & positive feedback loops
153 Archeological digital blueprinting
154 The Computer History Museum
156 The mysterious hole in M.C. Escher's Print Gallery
160 Who's talking math?
162 Mathematics uncovers the invisibility cloak
164 Mathematics, physics & art all in one-the art of Michael Burke
168 The Archimedes palimpsest
170 The Stomachion puzzle
173 Mathematical origami
177 There's gold in them thar internet searches
180 Mathematics fine tunes weather prediction
182 Quipu...numbers & words?
187 Solutions section
191 Index
195 About the author

Mathematics is more than numbers, more than solving problems and doing computations. Mathematics is a way of thinking. It deals with patterns and helps make connections. It's used to describe our dynamic environments, our economies, our political systems. It's a major player in the world of entertainment and in communication. It's used to describe and create music and art. Mathematics provides tools for the architect. The scientist could not function without it. Mathematics helps uncover the smallest particles, yet makes it possible to deal with the vastness of the universe and other infinite worlds. Mathematics is an art. To practice it, the mathematician looks at patterns, numbers and formulas, and finds new ways to apply old ideas while simultaneously creating new mathematical concepts. Mathematics is a fascinating world of ideas, and it touches almost every aspect of our lives.

What's a mathematical snippet? It may be an idea, a problem, a puzzle, a piece of art, a text bite—giving you a glimpse into the vast world of mathematical ideas. The snippet is just a portal of discovery, a mere snapshot of mathematics at work. I barely touched the surface of a mathematical concept, but hopefully captivate you enough to explore it further.

the disappearing columns of saint peter's square

Many famous attractions world wide have mathematical connections. St. Peter's Square in Vatican City is one such site. Although thousands of people visit it daily, only a small fraction of its visitors zero in on the mathematics of this famous piazza.

Its designer, Gian Lorenzo Bernini (1598-1680), used properties of a trapezoid, an ellipse, a circle and perspective to create a welcoming square which can comfortably accommodate over 300,000 people.

Completed in 1667, the oval shaped piazza is embraced by two curved *colonnades* consisting of four rows of columns. The two colonnades were erected on opposite ends of the

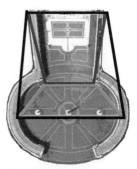

From a satellite view of *St. Peter's Square*. Note the two fountains straddling the obelisk in the center of the piazza. The colored segments indicate the various imaginary trapezoids.

Photo of *St. Peter's Square* taken while standing off to the side of the disk marked *centro del colonnaio*.

ellipse's major axis of 650 feet. Bernini used imaginary trapezoidal shapes within the piazza. The front of the Basilica and the ellipse's major axis (opposite parallel sides of the blue trapezoid) form one trapezoid. Within it are the green and red trapezoids. By creating these imaginary trapezoids the perspective one would naturally experience when entering a rectangular space becomes exaggerated and enhanced by the trapezoids and additionally directs the visitor's attention to the Basilica's facade.

One of the disks between the obelisk and the fountain which is the center for the imaginary four concentric circular arcs of the colonade's columns.

Within the elliptically shaped piazza two fountains are located on opposite sides of an obelisk which stands in the piazza's center. The fountains were positioned so that they are located at the focal points of the ellipse. Located between each focus and the obelisk are two disks. Stand at any location in the piazza other than on these two disks, and the front of columns and other rows of columns in colonnade are visible. However, if one stands on either of

Photos of *St. Peter's Square* taken while standing on the disk marked *centro del colonnaio.*

the disks marked with the phrase *centro del colonnaio* all the columns behind the front row disappear. It's both startling and intriguing. Illusion or mathematics?

What's the mathematics behind this vanishing act? It all has to do with the properties of a circle. Imagine standing at the center of four concentric circles. Shown here are four concentric arcs of columns. Radii emanating from the common center contain all four columns. If a person is at the circle's center, the columns on the line of radius are obscured by the front column on its radius. The viewer's line of perspective makes the other three columns behind the front one vanish behind that column.

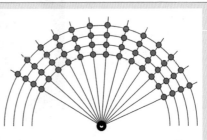

Imagine an elevation of this diagram, so that the gray spots represent columns of St. Peter's colonnade and you are standing at the black center of the four concentric circles. As you turn your head the columns within the span of your vision are only those in front row, while the columns in the back three rows will not be visible.

What about the ellipse: An *ellipse* is the locus of all points on a plane so that the sum of the distances from two fixed points, called *foci*, is constant. An ellipse has two axes of symmetry. The longest is called the *major axis* and the smallest is known as its *minor axis*. The closer the foci are to one another, the more the oval shape resembles a circle. When the two foci coincide the ellipse is a circle. In fact, a circle is a special case of an ellipse.

The *major axis* is segment AB.
The *minor axis* is segment DE.
C is the center of the ellipse.
The ellipse's area is $\pi|CB||CD|$.
If the axes of this ellipse coincide with the x & y axes of the Cartesian coordinate system, its equation is
where a=|CB| b=|CD|.
$$\frac{x^2}{a^2}+\frac{y^2}{b^2}=1$$

who really chooses whom—x, y or z?

This example illustrates how different voting systems can affect the outcome of an election.

Suppose candidates X, Y and Z have 35%, 32% and 33% of the votes respectively.

- In a *plurality election*, X wins because X got most of the votes.

- In a *run-off election*, Y is eliminated, and its votes can make either X or Z a winner in the run-off.

- In a *Borda election*, candidates are ranked and assigned points in order of preference. The first ranked candidate gets **2** points, the second **1**, and the third **0**. The winner is the candidate with the most points. Suppose 35% ranked X-Y-Z, 32% ranked Y-Z-X, and 33% ranked Z-Y-X. The points would stack up as follows:

Consider the percentages as a total of 100 votes cast and we get the following ranking:

X-Y-Z gives
   2(35) or 70 points to X
   1(35) or 35 points to Y
   0(35) to Z

Y-Z-X gives
   2(32) or 64 points to Y
   1(32) or 32 points to Z
   0(32) to X

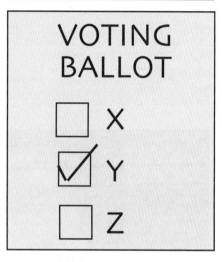

Z-Y-X gives
   2(33) or 66 points to Z
   1(33) or 33 to Y
   0(33) to X

Tallying the results, X got 70 points, Y got 132 points and Z got 98, so Y is the winner.

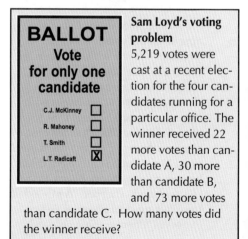

BALLOT
Vote for only one candidate

C.J. McKinney ☐
R. Mahoney ☐
T. Smith ☐
L.T. Radicaft ☒

Sam Loyd's voting problem
5,219 votes were cast at a recent election for the four candidates running for a particular office. The winner received 22 more votes than candidate A, 30 more than candidate B, and 73 more votes than candidate C. How many votes did the winner receive?

5

digital forensics

Today digitizing music, photos, movies, and videos has reduced sound and images to strings of 1s and 0s. A photograph or a video is no longer a smooth blend of colors flaws such as scars, wrinkles, marks can be removed— in other words *what we see or hear may not be what actually exists*. The old adage—*Pictures don't lie.*— no longer necessarily holds true.

but a conglomeration of pixels that appear on our computer or TV monitors—pixels which appear uniform to the naked eye, yet in reality are disjointed discrete dots. The digital revolution comes with its pros and cons. Music that was recorded in the past that may have acquired "noise" can now be cleaned up. Photos can be enhanced, colors become more vibrant, blurred pictures made sharper. A missed or incorrect musical note can be corrected. Photographic physical

Imagine every bit of data whether sound, image or text being converted to numbers — strings of 0s and 1s. Instead of a smooth seamless

The enlargement on the left brings the pixels of the photos into view. Each pixel on a color monitor is composed of up to three dots, a red, blue, and green that should merge to a point. The blend of these three colors determines the pixel's color. Each of these color dots uses 1 byte or 8 bits of data. Some monitors use only 1 byte, referred to as 8-bit. These screens are limited to 256 or 2^8 colors or shades of gray. A 24-bit display system uses all three bytes making it possible to show 2^{24} or 16,777,216 different colors or shades of gray.

stream of data, it is now composed of tens of millions of discrete packets of pixels or 1s and 0s that are distinctly associated with a particular sound, color, or hue. Digitizing translates a sound or image to its number equivalent. Here the binary and hexadecimal numbers rule. Here fractal geometry can transform huge amounts of data to a compact formula that can recreate the image or sound simply by plugging in a set of numbers to the fractal formula. It's not perfect or smooth, but seems perfect to the naked eye or the human ear.

Granted photos have been manipulated for decades, but not with the ease and complexity that today's computers and software allow. No darkrooms are necessary, only expertise of such programs as Photoshop and the computer facility to cut, copy, paste, click, drag and drop. Even today's scientific papers, reports, and evidence must be scrutinized for digital changes. For example, the 2004 news breaking story of South Korean scientist Woo-Suk Hwang's work on cloning human stem cells immediately lost its credibility and was withdrawn from the *Science* journal because of fraudulent data and photographs of human stem cells that had been digitally altered. Apparently the cells had not actually been cloned in the laboratory, but rather cloned digitally. *"The federal Office of Research Integrity has said that in 1990, less than 3% of allegations of fraud they investigated involved contested images. By 2001, it was 26%. And in 2006 it was 44.1%."*[1]

Digitizing has spawned a new set of problems and careers. With sophisticated high tech tools and computer technology it has become easier to alter, edit, delete or add information to a digitized document, photograph or sound. It is so sophisticated that it is extremely difficult for the naked eye to tell where fact becomes fiction. How does one distinguish what is original from the frauds & forgeries? By using mathematics as your spyglass and detective in the evolving new field of *digital forensics*. Digital forensic methods and programs such as Q by engineer Hany Farid are developing techniques to sniff out frauds. Farid's program can analyze the consistency of the lighting in a photograph, the distortions of pixels, the reflection of light and objects appearing in a person's eyes. Numbers also play an important role. The numerical value associated with the pixels of a digital image can be analyzed for traces of alterations. So although we can blame mathematics, namely binary numbers and computer innovations, for creating a realm of all new problems in forgeries, we can also look to math for the solutions to these problems.

---

[1] From *Proving That Seeing Shouldn't Always Be Believing* by Claudia Dreifus. *New York Times*, Oct. 2, 1007.

boids & birds

Whether it's schools of fish, flocks of birds, pods of whales, migrating Monarchs or herds of wildebeest, mathematics can be used to explain and describe movement and motion, predict paths, and even animate virtual animal movements and migrations.

Migrations and movements of animals and insects are a form of complex behavior influenced by such factors as reproduction, food supplies, predators, genetic codes, changing environmental conditions.

Chaos theory, nonlinearity, cellular automata, statistics and probability are used to describe and perhaps predict migratory behavior and movements.

new generations to evolve and interact with their surroundings. There are many mathematical concepts that come into play. Among these are measurement of distance and angles of the boids from the center of the group, dynamic equilibrium, and various algorithms and artificial life computer programs. Boids possess *emergent behavior*, that is, a boid can become (emerge) more complex as its individual components (in this case birds) interact following a set of rules. Examples of such rules are: (1) do not crowd adjacent birds, (2) fly toward the flock average heading, (3) try to stay within the average position of the flock. As a result the behavior (i.e. the movement) of boids is dynamic and can fluctuate between appearing chaotic or orderly.

In 1986 Craig Reynolds created generic simulated flocking creatures know as *boids*. Using computer simulation, he developed an *artificial life program* which is able to simulate flocking behavior of birds and movement of schools of fish. The interactive behavior of moving groups is *nonlinear*, meaning that the same conditions do not always result in the same outcomes. Reynolds found that boids can be described by *3-dimensional computational geometry* in which very small neighborhoods interacting with their surroundings influence the general overall reaction of flocks, schools or herds similar to how *cellular automata* depend on initial rules which allow

Zoologist Frank H. Heppner painstakingly filmed and studied flocks of birds, and concluded that birds were not guided by a leader, but flew in a state of dynamic equilibrium. Until he was introduced to chaos theory and computers, he was unable to successfully describe the flocking movement of birds. Heppner devised a computer program that uses four simple rules based on avian behavior and concepts of chaos theory. Moreover, the mathematics of computer modeling has been invaluable in simulating movements of boids. See websites http://www.red3d.com/cwr/boids/ ; http://www.vergenet.net/~conrad/boids/ ; and http://www.navgen.com/3d_boids/.

getting a glimpse of the mathematical world of salvador dali

Salvador Dali's surreal works appear as bizarre nightmares come to life. Disturbing multi-scene images of elongated figures, exploding flesh and contorted forms are sometimes intertwined with science and mathematics—concepts such as DNA in *Galacidalacides Oxyvibonucleaicacid* (1963), the warping of spacetime in *The Persistence Of Memory* (1931), non-Euclidean geometries in *Topological Contortion Of A Female Figure Becoming A Violoncello* (1983), and expressions of the 4th-dimension *In Searching For The Fourth Dimension* (1979). Dali's images are so eerie, one wonders if he was on hallucinagenics.

Photo of Salvador Dali. Courtesy of The Salvador Dali Museum. Detail from The Ecumenical Council, 1960. Copyright Salvador Dali Museum, St. Petersburg, Florida.

In his painting *Slave Market,* Dali created an oscillation optical illusion—our mind sometimes sees two women and other times Voltaire's bust. Courtesy of The Salvador Dali Museum. *Slave Market with Disappearing Bust of Voltaire.* (1940) Oil on canvas. 18 1/4 x 25 3/8 inches. Collection of The Salvador Dali Museum. St. Peterburg, Florida. ©2004 Gala-Salvador Dali Foundation, Figueres. (Artist Rights Society ARS) New York. ©2004 Salvador Dali Museum, Inc.

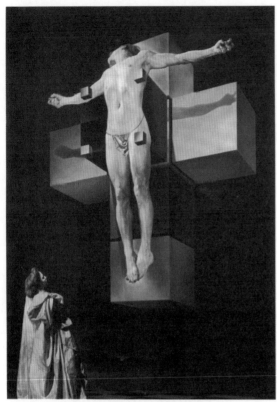

The Crucifixion (Corpus Hypercubus) (1954). Metropolitan Museum of Art. Gift of the Chester Dale Collection, 1955.

It's fascinating to see how Dali's used scientific and mathematical concepts in his art. In his *Palladio's Corridor of Dramatic Surprise* (1938) he relied on how parallel lines meet in infinity to make his painting zero in on a vanishing point. Dali was a master of optical illusions. His *Disappearing Bust of Voltaire* (1941) tricks our mind into shifting between two interpretations. In *Endless Enigma* (1938) our eyes discover and move between many hidden subjects, and in *Invisible Sleeping Woman, Horse, and Lion* his technique cleverly disguises these three figures.

Dali did not neglect exploring the concepts of four-dimensional space and the properties of non-Euclidean geometries. In *The Crucifixio (Corpus Hypercubus)* 1954, he used the unfolded hypercube (the 4th-dimensional cube that is unfolded into a cross composed of 8 cubes) with Christ suspended against it in space. Does Dali suggests that Christ was resurrected via the 4th-dimension? In addition, Dali depicts the hypercube's 2-dimensional shadow beneath the cross, created by a light source directly above the cross.

Yet Dali himself said *"Why should Dali use drugs when he has discovered that our world is a world of people with hallucinations, where theories, like that of relativity, add to the three dimensions of space a fourth, which is time, the most surrealist and the most hallucinatory of spatial dimensions. I have never taken drugs, since I am the drug. I don't talk about my hallucinations, I evoke them....I am the drug; take me, I am hallucinogenic![1]"*

Dali's ongoing fascination with mathematics continued throughout his life. In 1974 he visited with mathematician Thomas Banchoff of Brown University, who demonstrated how computers were used to explore the 4th-dimension. Dali realized how computers could be used to simulate and design stages of art. In 1979 he revisited the 4th dimension and painted *Searching for the Fourth Dimension*. In *Topological Contortion of a Female Figure Becoming a Violoncello* (1983), Dali used the concepts of topological transformations of stretching and pulling to depict the pain and motor spasms from which he suffered.

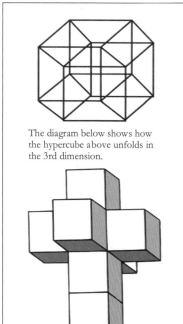

The diagram below shows how the hypercube above unfolds in the 3rd dimension.

Some of Dali's paintings also enter the world of physics. In *The Persistence of Memory* (1931), we see limp pliable clocks draped over objects. He thus portrays how time is manipulated as one of the four dimensions in Einstein's theory of relativity in which clocks, i.e. time, can be distorted when the clocks approach the speed of light. In *Soft Watches* (1933) pliable clocks again appear in the shadows along the sands of time, yet on the sand a key and a wilted rose lie in the light, perhaps signifying how the speed of light holds the key to time. In Dali's *Agnostic Symbol* (1932) an extremely elongated thin straight spoon appears emanating from the sky into dark space. The spoon then bends and curves around a mass as it passes near it. Is Dali showing us how the force of gravity caused by the curvature of space can bend light?

Whether it was simply a math term to name his work as *Singularities* (1935-36) or whether it became the subject as in *The Persistence of Memory*, his surreal art brought mathematics out from the subconscious level onto the canvas. Salvador Dali must be counted among the artists who consciously used and explored mathematics and science in his work.

---

[1] Dali by Dali; translated from the French by Eleanor R. Morse. Harry N. Abrams, Inc. Publishers. New York, 1970.

math or magic?

Many have been sent this magic trick via the internet. Even as you try it repeatedly, you are stunned by the seemingly magic results.

You're asked to follow these steps:
(1) Pick any two digit base ten number.
(2) Add together the two digits comprising the numbers.
(3) Subtract this sum from the original number you chose, and remember this result.
(4) Look at the table above, and find

will always be a multiple of nine less than 90.

How Does This Happen? The use of a little algebra will illustrate why the final result in step (3) is always a multiple of nine.

Step (1): Suppose you chose any two digit number. Call it **ab. a** is any digit greater than 0, and **b** is any digit less than or equal to 9. Since **a** is in the tens place and **b** is in the ones place, the value of number **ab** is $10a + b$.

95	84	73	62	51	40	29	18	
94	83	72	61	50	39	28	17	7
93	82	71	60	49	38	27	16	6
92	81	70	59	48	37	26	15	5
91	80	69	58	47	36	25	14	4
90	79	68	57	46	35	24	13	3
89	78	67	56	45	34	23	12	2
88	77	66	55	44	33	22	11	1
87	76	65	54	43	32	21	10	0
86	75	64	53	42	31	20	9	
85	74	63	52	41	30	19	8	

the symbol corresponding to your final result.

For this set of symbols your result is ♆. When you try it again on the internet with a new two digit number, a new set of symbols is presented. Once again it guesses the right symbol. In fact, every time you try, the computer gets the right symbol. WHY?

It's not magic, but tricky mathematics. Anytime you do the steps above, you will always end up with a final result that is a multiple of nine. Notice that all numbers listed above—and on any subsequent list generated on the internet—are multiples of nine less than 90 and have the same symbol. Why? Because your answer

Step (2): The sum of the number's digits is: $a + b$.

Step (3): Now we are asked to subtract this sum from the original number. So we get: $10a + b - (a + b)$ which equals $9a$. This means the final result is always a multiple of 9. The largest two digit number you can choose is 99, and its result is 81. In fact, no result will be larger that 81. Other numbers may also have this symbol, but that is done to throw you off track. You do not care which symbols are placed with multiples larger than 81 or which symbols are used for any of the other numbers since they will not be your result because they are not multiples of nine.

the alhambra &
the american institute of mathematics

The Alhambra in Granada, Spain.

Silicon Valley will be the location of the Spanish castle headquarters of the American Institute of Mathematics. The Institute's new site, whose projected date of completion is 2009, is the vision of John Fry, the founder of Fry's Electronics. Fry is modeling the headquarters upon the Alhambra in Granada, Spain. It will have geometric patterns and tessellations adorning the walls, ceilings, and columns and symmetries in its gardens and a copy of Alhambra's reflecting pool. The Silicon Valley Alhambra will be located on Fry's 190 acres in Morgan Hill, and will include underground parking and modern amenities in addition to replicas of such Alhambra treasures as the Fountain of Lions, and the fabulous tessellations that inspired M.C. Escher.

A Moorish tessellation gracing one of Alhambra's walls.

einstein's cosmological constant blunder?

Was Einstein's cosmological constant, *lambda,* his greatest blunder as he finally contended in 1931? Over the years Einstein was not afraid of making, admitting, and adjusting mathematical errors. As he himself said *"Do not worry about your difficulties in mathematics, I can assure you that mine are still greater."* While developing his general theory of relativity Einstein admitted to numerous mathematical and theoretical errors. In fact, he said of himself

"That fellow Einstein suits his convenience. Every year he retracts what he wrote the years before."[1]

In 1916 when Einstein applied his general theory of relativity to the universe, he was surprised to discover that his equations implied the universe was anything but static, but instead was either expanding or contracting. This contradicted past beliefs that the universe was unchanging. His results meant that the concept held by such great minds as Aristotle, Copernicus and Newton were wrong. Even though in the past Einstein often went against established scientific

ideas, here he couldn't bring himself to believe what his work had unveiled. Were his own innate beliefs too strong to accept his new results? Regardless, he was compelled to find a way to uphold the static universe concept. He introduced a constant, *lambda,* denoted by the Greek letter Λ, called the *"cosmological constant"*, into his equations. It would eventually adjust any contractions or expansions of the universe over time. The universe was now unstable but static. In 1922 mathematicians Alexander Friedmann proved that Einstein's equations still implied an expanding or contracting universe even with his lambda. Einstein was skeptical of Friedmann's results, but upon examining Friedmann's work he accepted the conclusion. Then in 1927 mathematician Georges Lemaitre found that if Einstein's theory was correct the universe could not be static. In 1929 Edwin Hubble produced evidence that the universe was expanding. In 1931 Einstein conceded that lambda was the greatest blunder of his career. Yet, in the last decade of the 20th century astrophysicists resurrected lambda and introduced it along with the new notions of dark matter, dark energy, omega, and the quantum effect. For now lambda is back in the 21st century explanations of the universe.

$$R_{\mu\nu} - \frac{1}{2}Rg_{\mu\nu} - \Lambda g_{\mu\nu} = \frac{8\pi G}{c^4} T_{\mu\nu}$$

Einstein's modified field equataion with the cosmological constant.

the big box for the smallest fabrications

Although nanotechnology deals with the world of ultra small fabrication, perched atop a hill in Berkeley, California is a huge rectangular box like structure overhanging its steep hillside by 50 feet. Six floors make-up the impressive structure. This safe deposit box-like structure is the home of the Molecular Foundry at Lawrence Berkeley National Laboratory. Rising four stories above ground level with two stories buried underground, the Molecular Foundry houses the latest laboratories for nanoscience. Since it's located about one mile from the Hayward Fault, the engineers at Rutherford and Chekene Engineering anchored the building with piers driven 100 feet into the ground and tied them together with poured concrete for added security. In addition, 12 foot high trusses run along the top of the building which both support the cantilevered floors and absorb vibrations from each floor. The entire exterior, including its trusses and rooftop mechanical systems, is covered with an aluminum paneled skin.

Rendition of the Molecular Foundry at Lawrence Berkeley National Laboratory.

Each floor of the building is designed to complement an area of nanotechnology and its labs. The top floor is designed as the Organic and Macromolecular synthesis; the second floor is for nanofabrication. The two bottom floors, insulated from minute vibrations and radiation, are nestled into the rock frame of the hillside. They house the electron microscopes which are able to write patterns on a single atom. The four floors above ground have their laboratories in the rear and offices and conference rooms in the cantilevered sections. Ecological consciousness is evident in the building's recycled aluminum facing, the recycled rubber used in its corridors, and its numerous energy conservation devices.

Big ideas in small packages are expected from this amazing structure.

dancing with the numbers

A dance performance may be the last place one expects to find mathematics and physics. Yet it's here we find choreographer Kathleen Hermesdorf's work *Fate+ Longing* inspired by string theory and mythology. Hermesdorf points out that although she is no expert on string theory, "*I feel like I understand it in my body. Dancers are scientists in some ways—They use, they observe, they fine-tune the world around them all at the same time. We all have to deal with matter in motion.*"[1] Her work has dancers moving and manipulating a stage full of golden strings. The actual strings in string theory are multi-dimensional and the minutest elements that determine the forms that matter and energy assume by their different vibrations. However, on stage their sizes are enormous and their effect on the dancers' movements profound. Hermesdorf's other work, *Blue 2000,* was based on the concept of an expanding universe.

Choreographer Erika Shuch's *Orbit* captures through dance the essence of slow zero-gravity movement, deals with the statistical improbability that a signal sent into outer space may never make contact, and with humans frustrating attempts to connect with alien worlds. Shuch explains "*I really wanted all that science to be in Orbit, but I had to decide which of the ideas really propelled the piece...the crazy thing about art and science (is that) our goal as artists is... to deepen knowledge.*"[1]

Thin Air by choreographer/dancer Donna Uchizono's turns to *quantum physics, the Heisenberg uncertainty principle* and the Buddhist concept of emptiness as the underlying forces which drive the dance. Before creating *Thin Air* she steeped herself in quantum theory in hopes of understanding some of its ideas and using them as springboards for

Whimsy and numbers with dancers on a Greek vase.

her choreography. Uchizono hopes her work conveys that *"the world as we know it doesn't exist in the way that we see it"*[1].

Whatever the artform, it's always fascinating to discover its connections with mathematics and the sciences.

---

[1] Quotes from *Dance takes a quantum leap into physics* by Mary Ellen Hunt. *San Francisco Chronicle,* Oct. 9, 2007

bridges that go anywhere

Grand Canyon Skywalk at Grand Canyon West

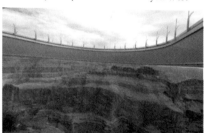

Bridges come in all sorts of forms, shapes and sizes whether one is dealing with a structure, a concept or a metaphor. We usually think of bridges being used to span opposite sides of physical gaps created by a canyon, a river or other natural occurrences. The *Skywalk,* a cantilever beam type bridge constructed at the edge of the rim of the Grand Canyon, connects two points on the same side of the crevasse. It's a walkway designed to give the thrill of being suspended 70 feet beyond the rim's edge and 4000 feet in mid air over canyon. Innovative engineering by Lochsa Engineering was crucial since the walkway does not only have to withstand the pull of gravity but is designed to withstand earthquakes and the 100mph vertical and horizontal winds of the Grand Canyon. Secured to the cliff with 94 steel rods bored 46 feet into limestone rock, it can support 70 tons of weight. Adding to the experience, the walkway's floor is clear glass that is 2.8 inches thick and composed of five layers of tempered glass. There are no supports above or below this u-shaped walkway.

Bridges can also connect apparently unrelated ideas or concepts. The wormhole, for example, can hypothetically connect different universes, dimensions, or times. And yes, wormholes are connected to mathematics. In fact, mathematician GeorgFriedrich Bernhard Riemann first discussed this topological bridge in his famous lecture of 1854. It was at this lecture that Riemann discussed ideas of a new non-Euclidean geometry—elliptic (aka spherical) geometry. He also spoke of n-dimensional space, wormholes and warped universes.

networking is not schmoozing

These two points and the segment connecting them form a mathematical system called a

network. The points are called the nodes of the system, and the segment is a link connecting the two nodes. Over the past two and a half centuries, from these simple elements, complex

mathematical ideas, expressions, theorems, definitions and formulas describing one of the most topical areas in mathematics, known as network theory, have evolved.

In 1736 Leonhard Euler drew a mathematical picture and used logic to solve the famous Könisgberg Bridge Problem. Reducing the bridge and land masses to the elements in this diagram was a stroke of genius that introduced mathematics to the world of topology and networks.

Among the mathematicians who have made contributions to network theory we find Euler, Paul Erdös, Afréd Rényi, Steven Strogatz, and Duncan Watts.[1]

Where is the mathematization of networks heading? To dynamic phenomena. The Königsberg Bridge

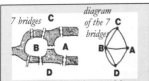

Königsberg Bridge Problem

Did a path exist which crossed each of Königsberg's seven bridges only once? Using the diagram above problem, Euler reasoned that a vertex could only have an odd number of paths *if* it was a beginning or ending vertex; otherwise, it had to be a vertex with an even number of paths through it. Since a path can have only one beginning and one ending point, there could not be more than two odd vertices. Thus, he concluded the Königsberg's problem was impossible because the diagram has four odd verices.

Problem and the properties of vertices and sides of polyhedron are examples of *static networks,* ones whose nodes and links are fixed—no bridge was added to the Königsberg problem, new islands did not surface, nor does the polyhedron grow an additional edge, face or vertex. On the other hand, *dynamic networks* constantly change. New links and nodes evolve adding to the complexity of the network and its problems. Dynamic networks include such systems as the network within a living cell, the network involving the links and nodes of the Internet, networks dealing with the spread of a disease, and even military networks composed of units engaged in combat. These networks require the mathematical properties of a complex system to describe them. They are neither completely regular nor completely random. They evolve in both a chaotic and orderly manner because a self-organizing phenomenon plays a major role in maintaining the delicate balance between order and chaos. A trucking transportation network is dynamic. The links of trucks and roads are constantly changed as are the nodes—the departure and destinations also can be altered. Yet this network and others, such as an economy based on supply and demand or the World Wide Web, do not solely comply with the rules of probability and random networks. Dynamic networks, especially involving

Topology is a field of mathematics which studies which characteristics of objects remain unchanged when they are stretched, pulled, scrunched as if they existed on a rubber sheet. In essence, it deals with transformations and how they affect the characteristics of objects.

Mathematically speaking, an example of **complete network** is penta- gon and its diagonals. The sides and diagonals make it possible for each vertex (node) of the pentagon to be directly connected to the other four vertices. Think of this pentagonal network as a group of five friends all of whom personally know one another. Such a network is complete. Now consider a circle with five points (nodes) on it. Each node is linked to the node immediately to its right and left. It's not a compete network, but each node can be acquainted (not directly connected) to every node via a number of other nodes.

 A **regular network** has each of its nodes with the same number of links, such as in lattice networks. Each and every node in the square lattice above has four links. The network in the honeycomb lattice below shows that each node has three links.

people or nature, are not governed solely by random acts. Most people do not make decisions just by flipping a coin, and nature is governed by very basic laws which minimize the expenditure of energy and materials.

Networks are everywhere. The flow of electrical impulses of the heart can be described by a network of synapses, the telephono system is a network which carries electrical signals along links consisting of wires, cables, fiber optics or radio signals, a person's connection to his/her great-great-great-aunt can be considered a genealogical network, and even the roots of a plant are a type of botanical network. The genome of your body and your nervous system are networks. Examining the flight paths of a particular airline in its flight magazine is another example of a network. Exports or imports of products is one of many examples of an economic network Everywhere we look we see networks connected to our lives.

---

[1] In 1959, mathematicians Paul Erdös and Afréd Rényi explored, analyzed and mathematized the properties of random networks. For a *random network,* there is no prescribed structure between the nodes and the links. Links can occur randomly so that every node has the same likelihood of getting a link although there is no guarantee.

• In 1998, mathematicians Steven Strogatz and Duncan Watts introduced, in their paper in *Nature,* networks referred to as *small world networks.* Small world networks are usually visualized in circular form which may range from regular networks to totally random ones. They were able to identify some of the properties of these networks in mathematical expressions. For example, in a network in which N nodes are connected to K neighbors, they found the maximum distance between two nodes to be N/K. They went further to prove that the distance between two randomly selected nodes is n/2K. They even randomly revised links between nodes in a regular network and derived the average distance between two of its nodes to be logN/logK. These findings have been applied to many real life networks such as electric power grids and the neural system of a nematode(C. elegans). It is felt that by focusing on small world networks, especially when probability and randomness are introduced, their properties will extend to networks in general.

• For in depth information on networks see *Linked* by Albert-László Barabási, Perseus Publishing, Cambridge, MA, 2002.

mathematics for math's sake

The 2006 Field Medals, the world's most prestigious math honor, were awarded to four mathematicians[1]. One of the medals went to Grigory Perelman for work that solved the century old Poincaré conjecture[2]. The conjecture involving topology and A 3-dimensional sphere was proposed in 1904 by Henri Poincaré, who inserted it as an incidental question at the end of a 65 page paper he had written. Experts feel the conjecture may be helpful in figuring out the shape of our universe. Unlike the other three recipients, Perelman did not show up in Madrid, Spain to accept his award. Apparently Perelman's interests lie only in knowledge not medals. As his former teacher Serge Rukskin said, *"Grigory is a devoted scientist in the pure sense of the word. He believes that the most important thing is that the problem is solved."[3]* In Perelman's own words, *"It (the award) is completely irrelevant for me. Everybody understood that if the proof is correct then no other recognition is needed."* [4]

Henri Poincaré stamp issued by France on Oct. 18, 1952.

Neither has Perelman expressed interest in the projected million dollar award that Clay Mathematics Institute will give for the solution to the Poincaré conjecture. Although Perelman's proof was not presented in a formal detailed paper, it has been scrutinized and found to indeed not only prove Poincaré conjecture but the more general *Thurston's conjecture,* in which Poincaré's conjecture is a special case.

If the Clay Institute announces an award, it will be sometime in 2008.

---

[1] Andrei Okounkov (Russia), Grigori Perelman (Russia) (declined), Terence Tao (Australia), Wendelin Werner (France) were the Field Medalists in 2006.

[2] Its proof received the 2006's *Breakthrough of the Year* award given by the journal *Science.* In topology your everyday sphere is considered a 2-dimensional object which exists in a 3-dimensional space because a sphere consists only of its surface which appears flat when we zero in on a portion of its surface. A sphere, a cube, the surface of a baseball bat are all the same objects to a topologist because they can be transformed into each other's shapes by pulling, scrunching and twisting. In topology what we consider a sphere is called a 2-sphere and it exists in 3-dimensions. On the other hand a 3-sphere is a higher dimensional sphere that exists in 4-dimensions. What the *Poincaré conjecture* proposed is that *the 3-sphere is the only possible finite shape for a 3-dimensional space that has no holes* (topologically described as finite and simply connected). In essence the conjecture describes the shape of our universe—here the Earth is a 2-sphere and sits in a finite 3-dimensional space which has no holes.

[3] *Reclusive Russian shuns major math prize* by Daniel Woolls. SF Chronicle August 23, 2006 .

[4] *Manifold Destiny* by Sylvia Nasar and David Gruber, The New Yorker. August 28, 2006.

zeno's "zany" paradoxes & more

Infinity and reality have had philosophers scratching their heads for thousands of years. In the 5th century BCE, philosopher Parmenides argued that reality was unchanging and absolute and that such things as motion and change were but illusions. These thoughts brought Parmenides under fire from fellow philosophers. For example, in Plato's dialogue *Parmenides*, Plato deals with Parmenides' ideas and questions the sensibility and seriousness of absolute reality. Because of the ridicule that was showered on Parmenides, it is believed Zeno (342?-270? BCE) was motivated to write his collection of paradoxes in defense of his teacher Parmenides. Zeno created at least 40 paradoxes, but unfortunately his notes were lost. These paradoxes are some of the most thought-provoking examples of what happens when one mixes infinity, time, motion, space, distance and reality. Fortunately, Zeno's paradoxes were preserved in the works of Aristotle and other philosophers and historians.[1] Whether Aristotle's renditions are true to Zeno's actual writings is unknown. In his account, Aristotle reflected his skepticism and tried to refute and dismiss their importance. Even so, Zeno's four most famous paradoxes—the *Dichotomy, Achilles and the tortoise, the Arrow,* and *the Stadium*

(Stade)—have intrigued for over two and half thousand years and continue to stir and baffle our minds today.

Even though Aristotle and others minimized Zeno's work, his paradoxes have withstood the test of time and have inspired others to write similar paradoxes. There are numerous spin-offs of Zeno's paradoxes. Here are two provocative ones:

Consider the *bouncing ball paradox* mentioned in G.J. Whitrow's 1961 book *Natural Philosophy of Time.* Here a particular ball is dropped and always rebounds 3/4 of its previous height. Since the rebound gets smaller and smaller, so does the time required to rebound. Using the ball's original velocity, Whitrow concludes that the ball bounces an infinite number of times in four seconds. How is this possible in reality? Does the ball never stop bouncing even though four seconds have elapsed? Do the ball's and the floor's imperfections and friction not allow it to follow the ideal theoretical circumstances? In theory, if it bounces an infinite number of times, what form do the bounces assume after four seconds have elapsed?

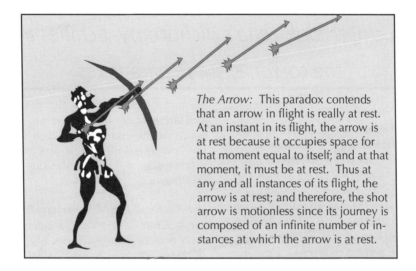

The Arrow: This paradox contends that an arrow in flight is really at rest. At an instant in its flight, the arrow is at rest because it occupies space for that moment equal to itself; and at that moment, it must be at rest. Thus at any and all instances of its flight, the arrow is at rest; and therefore, the shot arrow is motionless since its journey is composed of an infinite number of instances at which the arrow is at rest.

The Thomas Lamp is a fascinating paradox. Suppose a lamp has been rigged so that its on/off switch functions as follows when the lamp is activated: it stays on for 1/2 a minute, then goes off for 1/4 of a minute, then on for an 1/8 of a minute, then off for 1/16 of a minute, and continues doing this on/off scenario until the lamp's terminator is pressed. Suppose you press the terminator switch after one minute, is the lamp now on or off?

Paradoxes present circumstances which create contradictions, absurd results or astonishing questions while following a logical train of thought. They are often a two edged sword—both intriguing and disturbing. Over the centuries, Zeno's paradoxes stimulated the thinking of laypeople, philosophers and scientists touching on notions which dealt with space, time, motion, and the continuum. Mathematicians and scientists have explored and tried to explain his paradoxes by drawing on ideas from calculus and infinite series, the mathematics of infinity and transfinite numbers, and quantum physics and relativity theory. Regardless, loose ends remain; and, no doubt Zeno's gems will continue to spark our imagination.

Mathematically, a **continuum** is the set of points composing a line. This set is infinite and dense which means between any two points there's always another point.

a closer look zeno's dichotomy, achilles & the tortoise, stade paradoxes

The *Dichotomy:* Before an object can travel a given distance, it must pass through an infinite number of points before arriving at its destination. Consequently, a moving object passing through an infinite number of points cannot reach its destination. Consider an object traveling from point A to point B. First it must travel halfway to the point which is half the distance of A and B. Before it can get to the halfway mark, it must pass the 1/4 mark. Before reaching the 1/4 mark, it must get to the 1/8 mark. On and on, it must pass marks continually closer and closer to the starting point, an infinite number of marks. Therefore, it cannot get started, which means there is no motion—since it's impossible to pass through an infinite set of points, the beginning motion itself is not possible.

Achilles and the tortoise: In a race between Achilles and the tortoise, the tortoise is given a head start. But even though Achilles is much faster than the tortoise, this paradox contends that Achilles will never overtake the tortoise let alone reach it because by the time Achilles reaches where the tortoise was the tortoise has already gone ahead some small distance. So when Achilles reaches the end of each increment the tortoise moved,

the tortoise remains a bit ahead. On and on, there are infinitely many points Achilles must pass, which the tortoise has already passed; and therefore, he will never reach the tortoise.

The Stadium(Stade): This paradox involves relative motion. Consider three rows of three marchers, all identical. The middle row(the Ss) is stationary. The bottom row(the Ls) and top row(the Rs) move at exactly the same speeds but in opposite directions. Assume they are initially positioned as follows:

R1 R2 R3
  S1 S2 S3
    L1 L2 L3

Note that initially R3 and L1 are in line with one another. Then suppose the Ls move to the left and the Rs to the right until they are positioned as follows.

R1 R2 R3
S1 S2 S3
L1 L2 L3

R3 moves to the right only one space to be over S3 in an instant of time and L1 moves to the left only one space to be over S1 in an instant. But now L1 is two spaces away from R3. So how was it possible for them to be two spaces apart in a single instant of time? How is it that Ls and Rs move at the same rate with respect to Ss but twice as fast with respect to each other?

riemann hypothesis and more math
surfaces in case of lies

If you enjoy mystery novels you'll like *Case of Lies*. If you're a math buff, you'll love this book. It's chock full of mathematical concepts and ideas. Author Perri O'Shaughnessy (pseudonym for coauthors Pamela and Mary O'Shaughnessy) masterfully weaves mathematics, murder and intrigue while showcasing the *Riemann hypothesis*. The novel not only presents the math ideas in an engaging and understandable way, but also explains and seamlessly integrates them into the story line. We are introduced to the worlds of gaming, corporate security, the NSA and mathematical discoveries. Attorney/heroine Nina Reilly is drawn into a murder case and the world of mathematics and intrigue. Reilly is compelled to seek out and understand such ideas as division by zero, singularities, Euclidean geometry,the Pythagoreans, irrational numbers, prime numbers, the Riemann hypothesis, Cantor's continuum, Erdö's proof, Hermitian matrices, the *li* line, Gauss' prediction formula for the number of primes as one progressed along the number line and the Prime Number Theorem, Riemann's zeta function, and many more math concepts. Toward the end of the book "... *mathematicians seemed to be artists of form and number as surely as Picasso*

The *Riemann hypothesis* dates back to the mid 1800s, when Georg Friedrich Bernhard Riemann

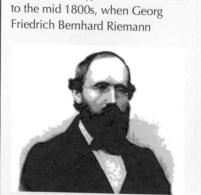

(1826-66) made his observation. During his life (1826-1866) he made many other phenomenal contributions to mathematics including developing a non-Euclidean geometry, proposing a 3-D universe that is warped in the 4th dimension, introducing the concept of wormholes. Among these he presented an ingenious hypothesis that has yet to be proven—*Riemann hypothesis*. It's connected to the zeta function— $\zeta(s)=1^s+(1/2)^s+(1/3)^s+...=0$, where s is a complex number. He noticed that the zeta function had no zeros (i.e. solutions) where the real part of s is ≥ 1. Where the real part of s is ≤ 0, the zeta function's zeros occur at negative even integers. These solutions are called *trivial zeros*. The amazing thing that Riemann noticed and hypothesized was that all remaining zeros, i.e. $\zeta(s)=0$, lie between 0 and

1. Riemann even went further with the hypothesis by claiming that the *zeta's function non-trivial zeros are only complex numbers with real part equal to 1/2.* In other words s=a+bi =1/2+bi. Moreover the zeta function is directly linked to the prime numbers through Euler's product formula. His hypothesis claims that the real part of s is always 1/2 whenever $\zeta(s)=0$.

$$\zeta(s) = \frac{1}{1^s} + \frac{1}{2^s} + \frac{1}{3^s} + \frac{1}{4^s} + \ldots = \sum_{n=1}^{\infty} \frac{1}{n^s}$$

Euler's product

$$\zeta(s) = \prod_{k=1}^{\infty} \frac{1}{1 - \frac{1}{p_k^s}}$$

p_k represents kth prime.

mathematicians have been trying to develop a formula or method for finding and predicting primes and factoring huge numbers. Pierre de Fermat, Martin Mersenne, Sophie Germain are among the many famous mathematicians who worked with primes. *"So much genius has been wasted trying to figure out what the hell primes are, and why they sit on a number line..."* (page 253, *Case of Lies*). In the search for finding and predicting prime numbers, the Riemann hypothesis has a prominent role even though it continues to elude mathematicians attempt to prove or disprove it. Regardless, mathematicians have developed and based new mathematical ideas on the assumption of its truth. Even physicists feel that energy levels of particles in quantum chaos are connected to the zeta function zeros. *Case of Lies* emphasizes the impact of prime numbers on our lives.

The Clay Institute of Cambridge MA lists the Riemann hypothesis as one of the millennium problems and will award a million dollars to the person who proves it.

was an artist of form and color. They were sensitive and jealous of their work like artists, too."(p.360 *Case of Lies*).

Today we are obsessed as never before with security and data protection, which are the underlying themes of *Case of Lies*. The global digital world relies heavily on prime numbers, which have fascinated mathematicians for over two millennia. Around 300 BCE, Euclid proved that there were infinitely many primes. Ever since then

polling & mathematics

Polling can be a tricky, yet profitable business. With billions of dollars going into determining what the public intentions are, guessing must not be part of the polling equation. Here's where mathematics and survey methodologies play crucial roles.

This is where *Bayesian probability* can be used to interpret information. Bayesian probability of an outcome relies on the degree to which a person or group of people believe that a proposition is true. *Bayes' theorem* can be used to create a model for rational judgement when only incomplete or uncertain information is available by using new information to shed light on the degree of belief. Bayes' theorem calculates *posterior probability* $P(A \mid R)$, that is the probability of A taking place with knowledge of a new piece of information, R. For example we know if a card is picked from a standard deck, the probability of it being an ace of hearts is 1/52. But if we now are told that the card drawn is red its probability is altered and denoted $P(A/R)$. Bayes' theorem has been applied to such fields as biology and creating spam filters.

Pollsters find that respondents to a poll tend to modify their answers, perhaps hoping to influence the outcome or to make their response fit within the realm of average respondents. Regardless of the reasons or how minor the given misinformation, these infringements on truth add up and can skew a poll's result. To counter such influence questions have to be devised that not only ask the respondents preferences, but what the respondents believe others prefer. The success of a poll lies in the care taken to formulate and phrase the questions in order to filter out and interpret answers correctly. Tweaking the survey will result in a successful poll or one totally off target.

Even though polling techniques and outcomes are improving, they are not perfect. Witness the New Hamsphire Primary of 2008. All nine polls predicated a significant lead for Barak Obama over Hillary Clinton, but the ultimate poll i.e. the primary election, had Hillary Clinton the victor. Polls and their predictions, like weather prediction, are not without flaws and misses.

Mathematician and minister, Thomas Bayes developed a special case of *Bayes' theorem* in the 18th century and was first published in 1764 by the Royal Society. In 1930s the idea of subjective form of probability was devised, and it was not until the 1950s the term *Baynesian* came into use.

---

juggling mathematics

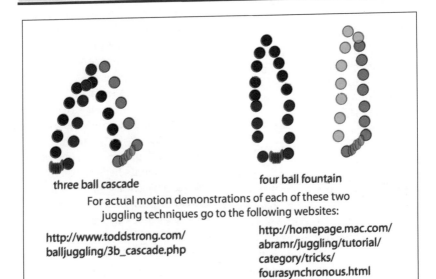

three ball cascade

four ball fountain

For actual motion demonstrations of each of these two
juggling techniques go to the following websites:

http://www.toddstrong.com/
balljuggling/3b_cascade.php

http://homepage.mac.com/
abramr/juggling/tutorial/
category/tricks/
fourasynchronous.html

We know juggling is no easy feat, and the mathematics behind juggling is even more complex. In the 1970s, mathematician and electrical engineer Claude Elwood Shannon (1916-2001), who is often referred to as the "father of information theory", came up with the first theorem[1] on juggling. In the subsequent decade, mathematicians began to look seriously at the patterns found in the art of juggling and applied higher mathematics to explain the science behind this art. Paul Klimek, Bengt Magnusson, Bruce Tiemann, Adam Chalcraft, Mike Day, and Colin Wright are among some of the mathematicians who began studying and applying mathematics to juggling.

Siteswaps, a mathematical notation devised to describe certain juggling routines, shows how patterns of juggling correspond to patterns of numbers. Siteswaps mathematizes juggling motion when two hands are alternately used and where two balls cannot be caught or thrown simultaneously. In siteswap notation, an *odd number* signifies throws that are passed between hands, while an *even number* signifies tosses caught by the same hand that tosses the object. Thus, a 6 could be described as a toss that stays in the air about 6 beats and is caught by the same hand. *The average of the numbers comprising a given sequence must always equal the number of balls being juggled.* For example, since the sequence

6-6-1-5-1-5 average is 4, four balls must be used to juggle this sequence following the beats indicated. As long as the beat is held consistently, its length does not matter. Thus, the siteswap 3-3-3 notation signifies 3 balls (since 3 is the average of 3-3-3) being tossed back and forth between the two hands at the same height in the air. The siteswap notation does not give precise information about the balls trajectory nor the movement of the hands. Nevertheless, siteswap notation has been invaluable for jugglers communicating and changing their routines. For example, the mathematics of averaging indicates that a 3-3-3 siteswap can be changed to a 4-4-1, since they both have the same average. A 4-4-1 has more variety in the height of the toss and the rhythm than a 3-3-3 siteswap.

Siteswap notation was only the beginning of integrating math and juggling. A few of the many fields of mathematics found today to describe juggling are: graph theory, combinatorics, Gaussian coefficients, permutations, topology, symmetric groups, and Galois num-

A juggler performing a four ball multiplex juggling routine using a 3 ball cascade pattern.

bers. It may have begun as a recreational problem and curiosity, but now the mathematics of juggling entails some high power mathematics and computer programming. Just do a google[2] search and see for yourself. WARNING: you may end up having a new hobby.

---

[1] Shannon's *Juggling Theorem* deals with a three-ball cascade. The exact equation is $(F+D)H=(V+D)N$, where F is the time a ball spends in the air, D is the time a ball spends in a hand, V is the time a hand is vacant, N is the number of balls juggled, and H is the number of hands.

[2] Some websites to check are: http://www.cix.co.uk/~solipsys/new/SiteSwap.html; http://www.juggling.org/papers/science-1/; *and* http://www.juggling.org/papers/science-1/mathematics.html

what's so special about random numbers?

Linking the words random and numbers seems contradictory.

Methods which have been used to generate random numbers include tossing coins, rolling dice, mechanical machines spewing numbers (as in bingo, lotto, and roulette wheels), and picking cards out of a hat.

An ancient Greek die on display at the National Archeological Museum, Athens, Greece.

up in the set of counting numbers {1,2,3,4,5,6,7,8,9,10,11,...}. In essence, a random number cannot be predicted, and a random number sequence must have no shorter way for expressing itself other than listing out the entire sequence.[1] In other words, there is no discernible pattern in the sequence. In 1927, the first table[2] of random numbers was compiled and published by statistician L.H.C. Tippett. His table consisted of 41,600 numbers. Yet random numbers go back thousands of years appearing first in games of chance as illustrated by the ancient die, one of the earliest random number generators.[3]

Numbers are usually associated with measurement, order and patterns while the term random implies fluctuation and lack of order. *What are random numbers?* According to the *American Heritage Dictionary*, the term random is defined as having no particular pattern or purpose: haphazard. It's an adjective linked to a spectrum of nouns from random acts and random thoughts to random numbers and fractals, random samples and random poles. We've heard the phrases "random acts" or "pick a number at random". Can a particular number be random in one instant and not in another? Consider the number 5. It may be a random number when it appears in the pips of a rolled die yet not when it shows

Today's global demands make past methods for generating random numbers far too slow and tedious. Random numbers are necessary for statistics and polling since they are used to select random and representative samples. Scientists use random numbers to model chaotic molecular behavior. Physicists simulate nuclear reactions with random numbers.[4] Without random numbers, gambling would not be gambling, and probability theory would be certainty theory. Even computer games would be totally boring if it were not for random numbers. Today in cryptography and encryption, there is a huge demand for these numbers. The Internet and government agencies are major consumers of random num-

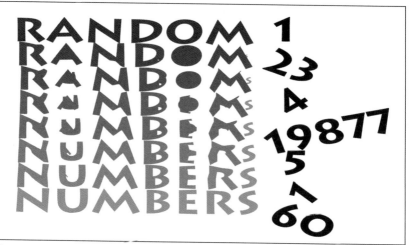

bers, using them to try to keep information secure. Hundreds and hundreds of random bits of data are needed every time one makes a single purchase by credit card through a secure server. Just do an Internet search for "random numbers" and notice all the websites which sell random numbers. Random numbers are big business.

Life and nature have many seemingly predictable outcomes. Yet predictable events may be altered with a random occurrence. For example, concentric circles are "always" created when a pebble is dropped into a pond, but a strong breeze can arise, altering the shape of the pebble's ripples. Random numbers are one of the ways mathematics connects to natural phenomena. Imagine a world devoid of uncertainty and unexpected changes. Such a world would have no chaos. Everything would follow a prescribed and predictable order. Monotony would replace change and excitement, and the mathematics of chaos would have no need for random numbers.

---

[1] For example, {1,4,3,16,5,36,7,64...} is not a random sequence because it can be reproduced by following the rule: To find the nth number of this sequence, square n if it's even, or just list it as n if it's odd.

[2] He determined his list of random numbers by tallying the middle digits of the measured areas of parishes in England.

[3] Dice were first used in ancient times by the Sumerians and then by the Egyptians. At that time, the numbers of chance created the idea of luck. Later on when probability became a mathematical subject, random numbers were used to measure odds and the predictability of an event. A random number generator is a tool or a means devised to produce strings of random numbers.

[4] Mathematician Stanislaw Ulam pioneered work in this area while working at Los Alamos, New Mexico.

the elusive black swan

The term black swan has its origins in ancient lore, when it was believed that all swans were white. In the 17th century, black swans were discovered in Australia and the term evolved to connote the

the term *black swan* to describe a rare unexpected, unpredictable event which has dramatic impact. Taleb explains *"Much of what happens in history comes from 'Black Swan dynamics', very large, sudden, and totally unpredictable 'outliers'...A black swan is an outlier,*

an event that lies beyond the realm of normal expectations. Most people expect all swans to be white because that's what their experience tells them; a black swan is by definition a surprise."[2] These surprises can be global events of historic proportion or even an event that has an historic affect on an individual's life.

possible existence of something impossible. 20th century philosopher Karl Popper used the black swan analogy to logically explain that just because all previous cases of something one is trying to prove continually have a particular outcome, does not prove the next case will have the same outcome.

The book *The Black Swan: The Impact Of The Highly Improbable*, by mathematical trader and writer Nassim Nicholas Taleb[1], applies

Why are black swans so elusive? Taleb points out that *"Our track record in predicting those events is dismal... We have a bad habit of finding laws in history (by fitting stories to events and detecting false patterns); we are drivers looking through the rear view mirror while convinced we are looking ahead."* [3] Examples of black swans include: September 11, 2001, the market crash of 1929, the amazing global impact of the Internet, the discovery of the light bulb— each of which set in motion a chain of events which affected the direction of history. Today, our global interconnectiveness and abilities to instantly communicate around the world can, as never before, quickly escalate the affects of a black swan.

Identifying a black swan that has happened is one thing, but according to Taleb we are not capable of predicting one. He explains humans are not genetically wired or conditioned to predict a black swan. Not only do our emotions get in the way, but so does our prehistoric genetic make-up. We have been conditioned to react to certain events, but not recognize and identify the infinite, the subtle minute events that are pointing in a direction we don't see. We are always looking for the immediate danger and not the ramifications of all the hidden possibilities. Taleb also claims that mathematical modeling has reached its limits, and even our enormous computer power capabilities are inadequate to cover the infinite possibilities of dynamic complex nonlinear systems. Yet are not Taleb's arguments creating a "black swan" as described by Karl Popper. Just because we have not been able to predict a black swan event thus far, does it prove we never will?

Although mathematics does not enable us to predict the future, it explains why events such as black swans occur. It's all linked to dynamic systems—systems that are constantly changing, constantly evolving, constantly affected by the large and minute ongoing occurrences. It exemplifies the famous *butterfly effect* and *strange attractors*. The black swan is the event that takes us over the edge of the delicate balance between order and chaos. The mathematics of non-linear complex dynamic systems explains why patterns are only predictable in a statistical sense, and even that is iffy. Mathematical models are just that, models. They can get "lucky" and be on target, or they can completely miss predicting a future black swan. Mathematics can provide us with tools to bridge what we're are missing in our genetic make-up and perhaps make it possible to foresee a black swan. Perhaps predicting black swans will be in the future realm of mathematics by enhancing its present tools— namely fuzzy logic, chaos theory, randomness and numbers, complexity, fractal geometry, automata models, power law, stochastic geometry, statistics, probability—with those that have yet to be discovered or evolved. But the remaining question is: Are we capable of accepting a prediction?

---

[1]Taleb's thought provoking books, *The Black Swan* and *Fooled by Randomness*, and his essays are very fascinating reads.

[2]From http://www.edge.org/3rd_culture/taleb04/taleb_index.html, interviews and exerpts.

[3]Ibid

voting paradoxes

Does the outcome of an election always reflect the majority's preference? Voting is no simple matter. The numerous systems of voting create disturbing paradoxes. In fact, sometimes the outcome of an election is determined by the system rather than by the votes cast or the voters' preference.

In the 1700s, mathematician and political philosopher Marquis de Condorcet discovered a voting paradox, now called the *Condorcet paradox*. It explains that majority voting is not a transitive operation. In other words, in an election it is possible for a majority to prefer A to B, a majority to prefer B to C, but this does not imply that a majority necessarily prefers A to C. In fact, the opposite is often the case.[1] In the 20th century, voting paradoxes were not given much consideration until economist Duncan Black wrote about this paradox in the 1940s. Stimulated by Black's work, economist Kenneth Arrow, a Nobel Prize winner, actually proved his impossibility theorem in 1951. In

Reproducions of voting disks used in ancient Greece.

his doctoral thesis, he showed that it was not possible to devise a voting system with a slate of three or more candidates that could select a winner and meet certain fairness criteria or axioms.[2] In 1952, mathematician Kenneth May proved that elections held using the majority rule involving just two candidates is the only voting system which is fair to both voters and the candidates.

Over the centuries numerous voting systems have been developed—all have been plagued by voting paradoxes. These include:

• *Plurality:* the winner is the candidate with the most number of votes, regardless of whether it is a majority or not.

• *Run-off election:* If no candidate receives a majority of the votes, a run-off election is held between the two candidates with the highest scores in the first election.

• *Sequential run-off election (or the exhaustive ballot):* If no candidate receives a majority of the votes in the first election, the candidate with the lowest score is eliminated. Then a new election is held. This process is continued until a remaining candidate attains a majority of votes.

• *Approval voting:* Voters are allowed to vote for all choices they find acceptable.

• *Condorcet election:* Here all possible pairs of candidates are matched up, and mini elections between each pair are held. The winner is supposed to be able to defeat every other candidate in a one-on-one mini election. Although the Marquis de Condorcet was a proponent of this system and strongly promoted its use, it had originally been proposed for an ecclesiastical election in the 13th century by Ramon Llull. The problem with the Condorcet process is that it can end without a winner.

• *Borda count:* Suppose a ballot is devised for five candidates. The voter ranks each candidate by assigning each candidate a point value selected from the whole numbers 0, 1, 2, 3, 4. Each point value can be assigned to only one candidate. In other words, the voter ranks the candidates in order of preference. The candidate with the highest number of points is the winner. Although 18th century mathematician Jean-Charles de Borda worked on and promoted this system, it was first considered in 1433 by German Cardinal Nicholas Cusa as a way to elect kings. Today the Bor-

da system is used in elections in Australia and Ireland and also to rank college athletic teams. Flaws in the Borda system surface when we consider that the voter is forced to rank candidates about whom he/she may be indifferent or unfamiliar.

Today mathematicians continue to study the voting paradoxes in hopes of devising or determining a system with the least number of flaws.

What areas of mathematics are involved in the voting process?
• When voters try to second guess how others will vote in a run-off or a sequential run-off election, you can be certain *game theory* enters the picture.[3]

• When an election is very close, any minuscule change in votes or campaigning strategies can have a major influence or impact on the outcome. Here *chaos theory* is at work. Consider the paradox of the single insignificant vote. In a large election, a single vote may carry little significance, but a group of these insignificant votes can affect the outcome.

• Sophisticated polling using *statistics* and *probability* influence how campaign managers design strategies.

• Today we find mathematicians such as Donald Saari putting mathematics to work to analyze the voting process with such tools as *chaos theory, algebraic symmetries, geometric struc-*

tures, bifurcation points, and *singularity theory.* His theorems, definitions and proofs shed new light on the dynamics of voting systems.[4] As Saari points out *"Voting is a basic tool of every democracy. But does the election outcome capture what the voters want? ... 'Not necessarily'. ...But why?! This has interested political sci-* *entists, economists, mathematicians, politicians and others for centuries. The source of these mysteries is now emerging, and mathematics and the symmetries it discovers are finally providing answers and resolutions."* [5]

---

[1]Suppose 30 people were asked to give their preference for A(apples), B (bananas) and C(cantaloupes). Now suppose when asked to vote between A and B, 20 preferred A and the remaining 10 preferred B. For their preference between B and C, 20 preferred B and 10 preferred C. But when asked about choosing between A and C, 25 chose C and 5 A.

[2]See Kenneth Arrow's book Social Choice and Individual Values for details.

[3]For example, consider the 1956 bill for federal aid for construction of schools which was introduced into the US House of Representatives. The House uses an amendment tournament method. The first vote is between the original bill, the amended bill, or no bill. If the amended bill passes, the run-off is between it and no bill. On the other hand, if the original bill passes, the run-off is between it and no bill. This method seeks to give influence to the second preference of a voter in case his/her first did not pass. Yet the process can be manipulated as it was believed to have been in this case. William Riker, renowned political science scholar, referred to as the founder of modern political science, developed a theory of how politicians maneuver issues for strategic advantage. In this particular bill, Riker points out that the Republicans did not want any form of the bill to pass. In the first round of voting, they deliberately sided with the Northern Democrats for the amended bill (giving federal aid to only desegregated schools) while the southern Democrats voted for the original bill (which gave federal aid regardless if the schools were segregated or not). The amended bill won this round. But in the second round, the Republicans now voted their first choice, no bill, along with the Southern Democrats, who preferred the original bill or no bill.

[4]Saari has proven that it is possible to design a voting system that selects a specific choice as long as one has an idea what voters think. A series of election packets are presented until the desired result is reached. For example, suppose 20 voters are split with the following preferences: 8 for ABC; 7 for BCA; and 5 for CAB. If you want A to be elected, you first have an election between B & C. This results in B being the winner with 15 (because B gets A's voters since B was their second choice). Then run B & A. A is the final winner with 13 (8+5). If you want B to win, first run A & C. C wins with 12. Then run B & C. B now is the final winner with 15. If you want C to win, first run A & B, so A wins with 13. Now run A & C. C is the final winner with 12. Saari's works include *Basic Geometry of Voting* published in 1994 by Springer-Verlag ; *A Chaotic Exploration of Aggregation Paradoxes* published in the March 1995 issue of SIAM Review; *Chaotic Elections! A Mathematician Looks at Voting,* AMS,Providence,RI.2001; *Decisions and Elections: Explaining the Unexpected,* Cambridge University Press,New York, 2001.

[5]*Mathematics & Democracy* by Donald Irving, lecture given at the University of California Irving, Oct. 24, 2000.

the 4-color map pr... quirky twist

E ven when a problem is ultimately solved, it may ... to rest. Consider the f... map problem—a ma...

... other two zigzagged
...en them making zigs and zags
... the size of the previous one. As
...ch country's boundaries
zigzagged among one another, their
borders became smaller and smaller

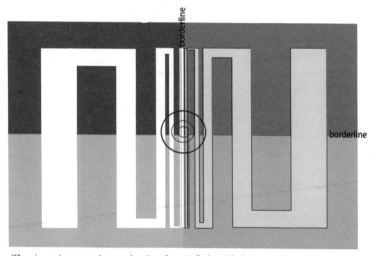

The zigzagging countries require six colors to distinguish them on this zigzagging map.

plane requires at most four colors to distinguish its boundaries. This problem emerged as a mathematical pastime in 1852 and spread throughout mathematical circles. It was finally proven in 1976 by Kenneth Appel and Wolfgan Haken, who devised a computer proof. Then in 2002, philosopher Hud Hudson created a plane rectangular island composed of six countries that zigzagged between one another, resembling a modern piece of geometric art. Four of its countries each occupied a corner,

and eventually converged to a borderline at the center of the island. It was shown that any circle drawn that includes this center borderline also has the six countries within it, regardless of how small the circle is. In other words, this center border belongs to all six countries, which means six colors are needed to color the island's map. Using this method, maps requiring any number of colors can be drawn up. Although this is a quirky twist, the four color proof still holds for conventional maps.

the road to chess

The origins of chess date back to 7th Century India with the board game *chaturanga*, a Sanskrit term referring to Indian war implements (horses, foot soldiers, elephants and chariots). From India the game was carried to Persia. From Persia it traveled along the trade routes to Russia and eventually into Constantinople. The Vikings brought it to Scandinavia, and it made its way to Europe through Spain with the Moors. As it moved through these countries it underwent changes reflecting the various cultures with which it came into contact. The names and shapes of pieces changed as did the way pieces were allowed to move. For example, the *king* was called the *shah* in Persia. The Indian *elephant* was replaced by the *horse* (called *faras*) by the Arabs and eventually became the *knight* in Medieval Europe. Initially a *pawn's* starting move was always restricted to only one square at a time. In Persia the *queen*—known as the *firz*, meaning counsellor— and her moves were limited until the end of the Middle Ages when the queen became the board's most powerful piece. It took about eight centuries to evolve to today's chess.

How chess pieces move today

The pawn moves only one square forward at a time, unless it is its starting move when it can either move one or two squares forward.

The knight can move to any of the dotted squares, and is the only piece that can jump others.

The bishop can move diagonally through as many squares, but cannot jump pieces.

The rook can move horizontally or vertically as many squares, but cannot jump pieces.

The queen can move as many squares in any direction, but cannot jump pieces.

The king can move only one square in any direction.

One would think that computers could be used to produce millions of random numbers. But computers use programs to produce data and solve problems, and their algorithms always follow a set of procedures. Even if a random number is used to seed (start) the process, the program will be inherently flawed for generating random numbers because it is functioning with a deterministic tool. The generated numbers may start out appearing random; but usually somewhere along the line, a pattern will emerge because of a program's repetitive procedure.* These computer-generated numbers, called *pseudo-random numbers,* are used as substitutions for random numbers, but they are not as safe or random as genuine random numbers. The only sure unbreakable method for encrypting information with random numbers is the one-time pad, which was mainly used during World War II. Here a sequence of random numbers is used to encode a message. That list of random numbers is only used for one message and then destroyed. The one-time pad method requires both the user and recipient to have possession of the list, thereby, making the process rather cumbersome. Currently mathematicians and computer scientists resort to entropy to produce the vast quantities of ranom numbers needed everyday. Sources of entropy can be produced by radioactive decay of matter, atmospheric noise such as in radio waves, electronic white noise and thermal noise.

An *entropic system* has inherent in it the absence of order. In other words, it is a system that undergoes spontaneous change that produces chaos and disorder making its results random. A random number generator must include a source of entropy.

Noise is any fluctuation or disturbance. It is not part of the intended signal and can interfere with the signal. For example, the website http://www.fourmilab.ch/hotbits/ uses an entropic system to generate random numbers by using radioactive decay to time successive particle pairs detected by a Geiger-Muller tube.

* Computers are deterministic tools with programs designed to produce the same result for the same inputted data. An early example of a pseudo-random number generator is John von Neumann's middle-square algorithm. Begin with a n-digit random number (the seed). First, square it. From its square, take out the middle n-digit number. (Note, if needed, use a 0 placeholder so that the n-digit number's middle digit can be selected.) This n-digit number is taken as the first random number. Then the process is repeated gathering a list of n-digit numbers that appears to be random. Eventually, a number is generated that is a repeat, and this number generates a loop repeating indefinitely a list of the same n-digit numbers previously generated. For example, suppose we start with a 3-digit random seed 461. Square it. We get 212521. Adding a 0 placeholder makes 0212521. The first 3-digit "random" number is 125.

Now repeat the process. Square 125 we get

15625 —>	562	0215296—>	152
0315844—>	158	23104—>	310
24964—>	496	96100—>	610
0246016—>	460	0372100—>	721
0211600—>	116	0519841—>	198
13456—>	345	39204—>	920
0119025—>	190	0846400—>	464
36100—>	610	0215296—>	152
0372100—>	721	23104—>	310
0519841—>	198	96100—>	610
39204—>	920		
0846400—>	464		

The number 610 appeared previously. Now all other generated numbers will be a repeat of the ones following the first 610.

the grand canyon & randomness

When awe conjure up an image of sand dunes or look at maple leaves, we see time and again how nature repeats shapes. Yet its shapes are never exactly the same whether it's the hexagon in a snowflake, the smooth recurring crests of waves, or the crags and cliffs forming the Grand Canyon. Some natural forms are easier to describe mathematically than others. For example, a sphere and its equation formulate the shapes of planets and the lune describes the crescent moon.[1] Some forms seem predictable and others utterly chaotic. But all are influenced by surrounding conditions—infinitely many conditions, which constantly change and affect the final outcomes.[2] Natural occurrences are dynamic systems. From a mathematical point of view, they are *complex nonlinear chaotic systems.*[3]

If the Grand Canyon had been designed solely by three-dimensional linear and quadratic equations, its diversity and awe inspiring formations would be lost. Straight neatly carved caverns and uniformly shaped parabolic mounds would replace its beautiful non-conforming layered formations. Instead of restrictions being dictated by mathematical objects, nature's formations are controlled by physical forces. For the Grand Canyon, these restrictions evolve from its environ-

The Grand Canyon. Arizona.

ment—the mineral composition of the soil, the constantly changing climatic conditions, the surrounding seismic occurrences, the impact of indigenous creatures, the flow of water, and global and cosmic changes. All the dynamic forces occur randomly. All of these can be simulated mathematically by using chaotic mathematical expressions. All rely on the use of random numbers.

---

[1] Other mathematical objects describing natural formations include the parabola for the path of a trajectile, an equiangular spiral for the orb spider's web, a mathematical meandering for the path of a river, and a random fractal for an ever changing coastline.

[2] Among these conditions are the availability of materials, interacting forces, and physical restrictions.

[3] In complex nonlinear chaotic systems, minute differences in a single minuscule condition among infinite initial conditions may create a huge difference in the final outcome.

the problem that determined gauss' career

This is less the story of the famous mathematician Karl Gauss (1777-1855), but more the story of a mathematics problem which determined the direction of his life. The construction of the regular 17-sided polygon is an easily stated prob-

an octagon, yet this problem's solution had eluded mathematicians for over 2000 years.

Then in 1798 Gauss, still a teenager, not only solved the 17-gon problem, but discovered much more. Gauss linked its geometric construction to an algebraic explanation, and unraveled its solution by conceptualizing the 17-gon on a complex number plane inscribed in a unit circle with its center at the graph's origin and placing one of the vertices of the 17-gon on the x-axis at (1,0).

The 10 Deutsche mark was issued on Oct.1, 1993 honoring Gauss' many achievements in mathematics and science.

lem—*using only a straightedge and a compass construct a regular 17-sided polygon (called a heptadecagon).* The ancient Greeks had figured out how to construct regular polygons with sides 3, 4, 5, 6, 15, and by simply bisecting the central angles of these polygons they also knew, therefore, how to construct all powers of 2s of the 3, 4, 5, 6, 15-sided polygons. Many people know how to construct a triangle, a square and a hexagon, but a heptadecagon is no simple feat. To the layperson it may not sound much different from making a hexagon or

From here he evolved and solved four quadratic equations, which enabled him to construct a segment of length $\cos(2\pi/17)$. This length meant he could construct an angle of $(2\pi/17)$ radians or $(360/17)°$, which is the size of a central angle of a regular heptadecagon. Yet, more importantly, Gauss *proved* that *regular polygons with a prime number of sides, p, can be constructed using a straightedge and compass if and only if p is of the form $2^{2n}+1$, where n is a natural number.* This theorem made it possible to know which regular polygons were constructible and which were not and that *any prime number resulting from this formula would be a constructible regular polygon.* Using Gauss' formula, $p=2^{2n}+1$ and letting n=0,1,2,3,4,5, we get that p=3; 5; 17; 257; 65,537; 4,294,967,297 respectively[1]. Notice the formula does not give the prime number 7, therefore proving that a 7-sided regular polygon is impossible to construct with only a compass and a straightedge. In addition, the formula also yields non-prime numbers, such as 4,294,967,297= 641x6,700,417. We see for n=2, the formula gives p=17 a prime number, and according to Gauss' theorem, the regular 17-gon must be constructible with the compass and straightedge, which Gauss performed. The 17-gon solution is an amazing accomplishment regardless of one's age, and bear in mind Gauss discovered more that its constructibility. His formula determines the constructibility of any n-sided polygon.

With the publication of this 17-gon work in the journal *Intelligenzblatt der allgemeinen Literaturzeitung* in June of 1796, Gauss immediately became known in mathematical circles. His work on the 17-gon problem was his first entry in his *Notizenjournal* — his mathematical/ science diary. Though Gauss lived for nearly 78 years, his diary consisted of only 19 pages with 145 subsequent entries.

Gauss considered his work on the 17-gon problem one of his greatest achievements, and requested the regular heptadecagon be placed on his tombstone. Unfortunately, legend has it that the mason did not comply saying it was too difficult and would end up looking like a circle. However, a 17 pointed star formed from a regular heptadecagon appears on a monument near Gauss' statue in Brunswick.

Problems are the catalysts of mathematical discoveries. They fuel the imaginations and curiosities of mathematicians. Their solutions often lead to new insights, sometimes to new problems and perhaps to the discovery of entirely new areas of mathematics. Until his 17-gon discovery Gauss was seriously considering studying philosophy or languages. For Gauss this problem pointed his career compass toward mathematics.

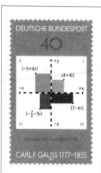

This stamp honoring Gauss' work with complex numbers was issued by Germany on April 14, 1977, to commemorate Gauss' 200th birthday anniversary.

---

[1]If one constructed a regular 65537-gon, it would be difficult to distinguish it from a circle. Even so, Harold S.M. Coxeter mentions in his book *Introduction to Geometry* that Johann Gustav Hermes constructed such a polygon in 1894 after working on it for over 10 years. His work on its construction is in the Mathematical Institute at the University Göttengen.

π's rock star status

Think about the irrational transcendental **pi** and its never ending non-repeating decimal. It is impossible to visually conceive the conglomeration of numbers that make up this infinite decimal number, yet it's amazing to see it

3.1415926535897932384662643
3832795028841971669399375
1058209749445923078816406
2862089936280348825342117
0679821480865132823066647
0938446095505822831725359
4081284811174502841027 ...

simply and elegantly described by *the circumference divided by its diameter.* Today pi has reached and perhaps surpassed rock star status with its own t-shirts, mugs, stickers, posters, post-its and even with an annual Pi Day...March 14th. The date and time written as **3/14** at **1:59am** expresses pi out to 5 decimals places, 3.14159. On March 14, many schools around the world pay tribute to pi with parties and games.

Other pi annual days that celebrate the *approximation of pi* include:

• July 22—22/7 the lower bound that Archimedes used for approximating pi.

• April 26—the day when the Earth's orbit divided by the time it has traveled thus far approximates its circumference/diameter. For leap years it's April 25th.

• the 314th day of the year, which falls on November 14 for a non-leap year.

• 355/113 was an early Chinese approximation for pi and is celebrated on December 21 at 1:13pm (355th day/113 time).

The simplicity of pi's definition versus the complexity of its number representation is so captivating that many people seem to be continually enticed to tax their minds and memories to capture pi's decimals out to more and more places. The latest record is held by Japanese mental health counselor Akira Harguchi, who recited pi to 100,000 decimal places in 16 hours, breaking his own personal record of 83,431 digits.

A *transcendental number* is a number which cannot be a solution of a polynomial equation with rational coefficients. $2x^2+5x-3=0$ is an example of a 2nd degree polynomial equation with rational coefficients 2, 5, and -3. This polynomial equation's solutions are 1/2 and -3. In 1882 mathematician Carl Lindeman proved π was a transcendental number by showing it could not be the solution of a polynomial equation with rational coefficients.

Ever wonder what happens when you google or search the internet for something? Consider the travels of such a *request*. Type the *data request* into your search engine. Click your mouse, and off it goes on its journey. Its first stop is your

webpages are ranked and a procedure is performed that arranges these web pages by listing and ordering them, usually by popularity. To accomplish this an equation is solved that may involve over half a billion variables. Then little snippet phrases are generated

web server, which probably consists of a network of over ten thousand computers communicating via the Internet. Your *data request* is then sent to the index servers which search their indices for web pages that match this *data request* and determine on which document servers these web pages are stored. From there it's sent onto the document servers which will retrieve the web pages that have your *data request*. Here these

which describe each web page listed with your *data request*. Finally the lists of the webpages with your *data request* appear on your monitor in groups of 10 or more. How long was the life of this search? Depending on your internet connection, it could be less than a second!

geoengineering

Geoengineering is the science that seeks to modify the Earth's environment on a grand scale in order to continue the Earth's habitability for humans. It's a relatively new science whose origins date back to the 1960s when scientists first considered the possibility that anthropogenic climate changes might be a future threat to humans.

In subsequent decades, when global temperature began to rise, some scientists began exploring ways to counteract climate change. Seemingly far fetched ideas and fixes have been proposed to counter or at best delay such changes.

> Some proposed geoengineering ideas are:
> (1) to create a shade in the stratosphere by placing a layer of sulfur dioxide, thereby shielding the Earth and reflecting more sunlight back into space to diminish the amounts of heat trapping gases.
> (2) to reflect more sunlight into space rather than have the Earth absorb it by: (a) placing lightweight (mylar type) mirrors in orbit to reflect sunlight before it reaches the Earth; (b) laying reflective film over large expanses of land such as on deserts or creating huge white synthetic plastic islands to reflect sunlight as they float on the Earth's oceans
> (3) to fertilize the oceans with iron in order to encourage the growth of vast amounts of sea vegetation that would take in tons of carbon dioxide and sink into the sea as they died off.

Mathematics plays a two edged role in geoengineering. It provides the tools such as complex computations and computer modeling to geoengineers for realizing their ideas. Yet it also points out the hazards of tinkering with dynamic non-linear complex systems such as those inherent in climate and ecosystems and their interconnections. The butterfly effect warns that a dynamic complex non-linear system, such as weather, may have chaotic outcomes from a minute insignificant events. Climatologist Ken Caldeira at Stanford University points out *"If you go down the path of geoengineering, it leads to taking ever increasing environmental risk, and eventually, you'll be unlucky."* [1] On a small scale we have seen ecosystem interference backfire time and again. For example, consider the one-hundred European starlings introduced to New York in the 1890s by an industrialist who wanted to introduce all the birds mentioned in Shakespeare's work. Today the starlings are a menace to native birds and farmers' crops. The failed effort to control the rat population of Hawaii by introducing the mongoose in 1883, has had devastating impact on native species.

Will geoengineering change our focus and commitment on vigorously seeking alternative green energy sources, reducing energy consumption, and cutting greenhouse gases? In case efforts fail to curb greenhouse gases sufficiently and alternative power sources cannot meet demands, should we keep options open and pursue geoengineering's innovative ideas?

---

[1] *Rebooting the Ecosystem,* David Wolman. WIRED magazine, December 2006.

the eyes that launched
the hearts of countless readers

In June of 1985 the face of a young Afghan girl appeared on the cover of National Geographic. The photographer, Steve McCurry, shot the photo at a refugee camp in Pakistan. The photo brilliantly captured the girl's haunting eyes reflecting the tragedies that beset her homeland. She was identified at that time only as the "Afghan

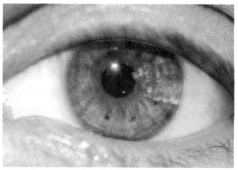

Photograph showing the intricacies of the eye's iris.

girl". The photo was subsequently used by Amnesty International in depicting the plight of refugees. Seventeen years after McCurry took the photo, he returned to the same refugee camp with a National Geographic crew in hopes of discovering what had become of the unnamed Afghan girl. Following various leads, McCurry came upon a young woman around 28 years old, whose eyes immediately told McCurry this was the person in his photograph. Her name is Sharbat Gula. He was able to photograph and interview Sharbat and her family. To verify that this was one and the same person, McCurry consulted two specialists in biometrics, the science that uses mathematical analysis to recognize and mathematize facial characteristics. Thomas Musheno, an FBI forensic examiner doing geometric facial comparisons, and John Daugman, professor of computer science at Cambridge University in England, using Daugman's iris recognition software were able to mathematically confirm that the Afghan girl and Sharbat were the same person. To view the actual photos and see for yourself go to *http://ngm.national geographic.com/2002/04/afghan-girl/index-text.*

The photos of Sharbat as an Afghan girl and as a woman were used to produce surface scans. These scans showed that the iris patterns of both matched. The biometrics of the eye has up to 400 measurable features. These features include the retina's blood vessel patterns, the iris' pigment granules, fibrous and vascular tissues and subtle movements. A biometric of the eye is done using an infrared light which scans the retina and iris. The results produce a digital barcode of the eye. Thus far the eye biometrics is the most reliable detection tool, barring the use of DNA biometrics.

the leonardo da vinci museum

If you ever wondered what da Vinci's sketches of his inventions would actually look like, the *Leonardo da Vinci Museum* in Florence, Italy is the place to visit.

Each display shows Leonardo's sketches and illustrates and explains the scientific and mathematical principles behind it.

Among the many exciting displays are:

One of the rooms at the Leonardo da Vinci museum in Florence, Italy. In the right foreground is Leonardo's machine gun. The stucture in the back is his armored tank.

It's a museum that lets you peek into da Vinci's mind. Here one finds life size models of most of Leonardo's scientific inventions. It's a hands on museum— you can touch and try most of the displays.

• *automatons* which include a lion, theater machine that changes scenes in a play, an odometer, a rolling mill, a column riser, a revolving crane, a printing machine,an olive oil press

48

- *water inventions:* a hydraulic saw, an Archimedean screw, floats for walking on water, webbed gloves, hydrometer for measuring the humidity in the air

- *air inventions:* a parachute, a bicycle ornithopter, a vertical ornithopter, wings, anemometer and anemomoscope

- *military inventions:* a machine gun, an armoured tank, a cannon

- *mechanical inventions:* a mechanism that transforms vertical and horizontal motions for moving heavy loads; a machine that converts rotational motion into vertical motion through a cranking mechanism; a machine using lantern pinions, a worm screw tooth crank which turns a wheel for lifting loads, a chain used with notched wooden gears; a locking system for hoisting heavy loads, a cranking flywheel system, ball bearings used in various devices to reduce friction, pulley and rope devices, a hammering device using a modified cam.

This device converts rotational motion into vertical motion.

A mechanism for producing concave lens or parabolic mirrors which could be used for boiling water by concentrating the Sun's rays on large vats of water. Its crank simultaneously spins two stone disks on two different axes which smooth a mirror located on the horizontal disk. Below is a portion of Leonardo's sketch of this device.

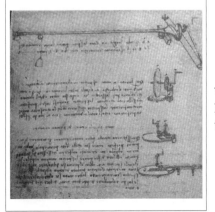

Visit the museum's website at http://www.leonet.it/comuni/vinci/ to view the complete collection.

to be or not to be a rembrandt?

Why are Rembrandts so iffy? During Rembrandt's career some forty pupils studied at his studio. During his lifetime many Rembrandt-like paintings were produced by his students, and some of these paintings were purchased

by art dealers, and passed off as original Rembrandts. Today there are dozens of fake Rembrandts both in private collections and in museums. Rembrandt experts are finding it difficult to agree on which of the speculative Rembrandts are authentic and which are not. Much is at risk, both monetary and artwise. The Metropolitan Museum of Art[1] has 40 Rembrandts only twelve of which have been authenticated as actual Rembrandts.

Because most of the fake Rembrandt paintings were done by his students, the canvas was prepared as Rembrandt had prepared his, and the wood and pigments were of that period. Even though experts have analyzed brush strokes and art techniques, certainty is questionable. Therefore, high tech methods such as x-rays, pigment analysis, dating methods, autoradiography, infrared imaging, and dendrochronology have their limitations when it comes to the works of Rembrandt.

Authentication has a lot at stake for collectors and museums. The some 700 works originally identified as Rembrandts have dwindled significantly over the past century as they have been placed under the scrutiny of experts and scientific techniques. Here the science of *stylometry*—analyzing, identifying, and quantifying the idiosyncrasies of writers' and artists' works — is also being put to work. The term stylometry was originally coined in 1897 by philosophy historian Wincenty Lutaslowski, and involves many branches of mathematics to analyze both literary and visual art works. For example, statistical techniques applied to word usage and style are used to determine authorship. In visual stylometry, wave theory and wavelets have been used to dissect a piece of art that has been scanned and pizelized.

Mathematician Daniel Rockmore has written software that reduces ultra high resolution photos of a painting to pixels, each of which is analyzed and assigned a numerical grayscale value which lies between the numbers 0(black) and 255 (white). Gradually an artist's brush strokes are peeled away leaving a blurred image that might help identify a particular artist's style by statistically analyzing each pixel. The pixel data is reduced to a point which is graphed on a multi-dimensional grid. Theoretically other works' points by the same artist should cluster. Rockmore first tested his software on the works of Pieter Bruegel. Thirteen drawing were analyzed. Its results plotted points in space in which all eight works of Bruegel clustered around one region of space, while the fakes were nowhere near Bruegel's. Whether the software will work for Rembrandt remains to be seen. As Rockmore points out *"it is likely that different media and different artists will require the use of different tools. ...this(visual stylometry) is in its infancy...It is only through the introduction of metrics and robustness analyses that a true science of visual stylometry can be created."* [2]

Wavelets can be used to describe anything that appears as a picture, sound, vibration or other form of data. A *wavelet* is a certain type of mathematical function which divides a given function or continuous-time signal into different frequency components so that each component can be studied with a resolution that matches its scale. The time component of a wavelet is called a *window*. Its size can vary according to the scale or resolution being used. When a combination of wavelets is used to describe a function, it is known as a *wavelet transform*. The wavelet transform converts a wave or signal into a series of wavelets in which each wavelet preserve its characteristics, while describing the wave or signal being transformed into its small wavelet components. Wavelet transforms have many applications and are used in various fields. Among these are: physics, astrophysics, seismic geophysics, optics, turbulence, quantum mechanics, in image processing, blood-pressure, heart-rate, ECG analyses, DNA analysis, protein analysis, climatology,fingerprinting, general signal processing, speech recognition, computer graphics, multifractal analysis, and image processing.

Even though wave theory dates back to the early 1800s with the work of French mathematician Jean Baptiste Fourier, a wavelet is a relatively new mathematical concept.

---

[1] In 1995 the Metropolitan Museum of Art had an exhibit *Rembrandt/Not Rembrandt* that brought together a collection of those authenticated as Rembrandts and those not. The museum goer could make their guesses, and read the expert analyses as to authenticity.

[2] From *Steps Toward Digital Authentication* by Daniel Rockmore and Greg Leibon.

how bees and other insects fly

In 1934 two European scientists, Antoine Magnan and Andre Saint-Lague used mathematics and the then known principles of aerodynamics and showed that bees cannot theoretically fly, even though we see them buzzing all over the place. Today using advanced mathematical modeling, scientists are discovering how bees and other flying insect actually do fly. Given the shape of the bee's body and its flexible wings it is not possible to compare the way a bee flies to how a plane gets off the ground. Airplanes, with their specially curved rigid wings, make it possible to balance the vertical forces of gravity and lift and the horizontal forces of

drag and thrust. On the other hand, an insect's wings are flexible, and in addition, its flight also relies on microscopic complex physical forces and effects that are useless to aircraft. Studies, computer models, and even robotic insects models have shown *"What really matters is not the shape of the wing, but how the insect moves it."*[1] The wings of insects move similarly to propellers in that they are constantly in motion. In fact the honey bee flaps its wings at about 240 times per second while the fruit fly's goes about 200 times per second. Computer modeling of this action reveals the formation of micro-tornadoes that form and remain at the upper end of the wings, called leading-edge vortices. This makes the air pressure within the tornadoes lower than the air pressure surrounding them. So the higher air pressure under the insect's wings creates the necessary lift to allow it to fly. The insect also gets upward movement from the many complex rotational movements of its wings such as backward and curling flaps. How the insect's minute brain is able to make continual adjustments to its movement while buzzing hither and thither remains a mystery.

---

[1] Dr. Dickinson from *Physics of flying keep insects as busy as a bee while in the air.* By Keaty Davidson, San Francisco Chronicle, Nov. 28, 2005.

Imagine the Universe with an infinite number of universes— ever increasing in number. History is full of such scientific theories. At one point atoms were the building blocks of matter, now science has discovered smaller subatomic particles, and it's believed the vibrations of minuscule strings may be shown to be the essence of matter. Yet in science, ideas are not accepted because they *seem* to fit perfectly in the scheme of things observed or because they sound brilliant. No, they become accepted if scientists devise methods to test them or have definitive information or results that point in the idea's direction. For example, around 140 AD Ptolemy formulated his theory of celestial bodies from things that had been observed. He proposed that the then known planets in our solar system traveled in circular orbits around the Earth while the Earth stayed motionless. His ideas were accepted as truth for centuries. About 1400 years later Copernicus challenged Ptolemy's geocentric theory with new ideas which claimed that the Earth and the other known planets actually orbited the Sun. Within a few decades the works of many scientists, among them Johannes Kepler and Galileo, came up with concrete evidence that upheld a heliocentric concept, but also showed the orbits were not

Above: *Snowflakes Cluster* abounds with strange shapes and stars in interstellar night. Courtesy of NASA, JPL-Caltech, P. S. Teixeira (CfA)

Below: A computer simulated cluster of galaxies. The large central region is an elliptical galaxy. Courtesy of NASA, ACS Team, Rychard Bouwens (UCO/Lick Obs.)

circular but elliptical. In addition, the classical and biblical concept that the heavens were *changeless* was refuted. Brahe Tyco's observations of a new star in 1572 and a comet in 1577 showed that the Universe does change. The work of these scientists proved what actual-

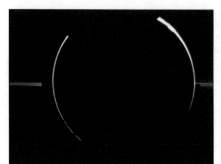

Jupiter's rings were first discovered in 1979 when Voyager 1 spacecraft passed by Jupiter. These rings remained a mystery until the Galileo spacecraft orbited Jupiter between 1995-2003. Scientists determined they were created by meteorite strikes on its nearby moon. These impacts exploded dirt and dust into Jupiter's orbit. The above picture shows the dust particles reflecting light caused by an eclipse of the Sun by Jupiter
Courtesy of NASA, JPL, Galileo Project, (NOAO), J. Burns (Cornell) et al.

ly existed as opposed to what seemed apparent. This knowledge was made possible by developing methods, experiments and tools (e.g. telescope) to actually observe and record phenomena that substantiated or refuted various theories. This is the way science works. This is its legacy.

Explanations by the ancients relied heavily on the imagination. In those times cosmology was connected to mythology. For example, the Big Dipper known as Ursa Major (the Big Bear) landed in the night sky because of Zeus' love for Callisto. The other constellations had equally fascinating names and stories of their origins. Since then our notion of the Universe has evolved as our

knowledge was enriched. There have always been far out ideas trying to make *sense and order* out of the cosmos. For example, Kepler felt our solar system was neatly connected to the five Platonic solids— the tetrahedron, cube, octahedron, and icosahedron. But new evidence, specifically the existence of more than six planets, demonstrated his idea was false. In the late 1800s mathematicians discovered new worlds in non-Euclidean geometries. Some mathematicians developed models that described worlds where Euclid's 5th postulate does not hold true. One such world was Henri Poincare's *hyperbolic world* in which everything—people, object, matter, forces— shrunk as they approached its boundary and grew as they neared the center. Since they shrank, this world did not appear finite to its inhabitants because it seemed to have no boundary or ends which could be reached. Many other "strange" things took place in Poincaré's world. For example, here, unlike Euclidean geometry, the shortest distance between two points was one that curved toward the center of the circle because steps became larger as one approached the center. There is no doubt that the quest for insights into the Universe will continue to stir the imagination.

mathematics behind the perfect golf ball

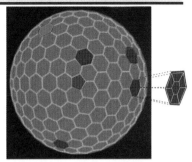

Calloway Golf had their engineers perfect their HEX ball's aerodynamics, which use 220 hexagons and 12 pentagons (see red pentagon). Each polygon is incised with a similar smaller polygon within it.

The everyday golf ball may look like a sphere, but it's much more. Using mathematics of optimization, including linear algebra, Calloway Golf has reinvented the conventional golf ball, changing it from the dimpled surface which was designed by English engineer William Taylor in 1908. Although the dimple design at that time had significantly reduced the drag of the guttie, a smooth surfaced golf ball of the 1800s, the engineers at Calloway came up with a lattice-type network with 220 hexagons and 12 pentagons, described as a hexagonal tubular lattice network. This lattice of hexagons and pentagons creates a turbulent boundary layer which affects the airflow surrounding the ball in flight and reduces its drag further. Techniques of *optimization*, mathematical modeling and testing have helped create a nearly ideal depth indentation for the hexagons and pentagons— a mere 0.0083 inches.

If you're tired of trying to locate lost golf balls, check out the new golf balls with ID tags which emit signals which lets a hand held device tell you if you are getting close. These balls are called RFID (radio frequency ID).

Optimization is a type of mathematical programming for finding the best(maximum or minimum) solution for a problem that has many possible solutions, like in the case of the best golf ball design to pursue. The optimization problems usually have many given conditions. For example, x numbers of dollars are available to produce y widgets. Various types of materials are available, each of which produces a widget with different characteristics and cost. Using linear algebra, multi-variable equations/inequalities for each set of conditions/ restrictions are written. The simplest problems can use two-dimensional graphs to solve the system of equations/ inequalities. The solutions of each equation/ inequality covers an region of a 2-D plane's graph. Where these regions intersect lie the solutions that satisfy the system of equations/ inequalities. If cost and performance of a widget is being optimized, one seeks the solutions on the graph that satisfies these two factors. With a multi-variable equations/inequalities, the problem can get hairy, and here is where mathematical modeling and computer analysis enters the picture.

mathematics caps off the dome

A dome[1] is a very simple yet elegantly shaped structure. Modern materials, tools and engineering allow today's architects to design not only traditional hemisphere structures but domes which open and close, domes held up by air pressure, and domes formed by a lattice of triangles.

other small structures. But the physics and mathematics behind these hemisphere structures had not yet been mastered let alone developed. One of the keys to how a dome might be constructed lay with the invention of the Roman arch[2], the 2-dimensional counterpart of the dome. Using their knowledge of the arch, Romans were the first to

Cathedral of Santa Maria del Fiore at dusk.

From mathematics we learn how the sphere encloses the greatest volume with the least expenditure of materials. A hemisphere likewise encloses the largest amount of space while using less materials than any other roof-shaped structure. Domes had their early origins as mounds of earth, as seen in ancient burial mounds. Later solid domes were used to top-off huts or

master the building of large masonry scaled domes, as witness the *Pantheon* in Rome, Italy, whose enormous dome tops a cylindrical building.

What has always made the dome a special architectural structure is its allusion to a heavenly canopy and its freedom from interior supports or columns. The stresses/thrusts exerted from the dome around its circumference are supported and absorbed by the cylindrical base. Although the structure's floor area is not circular, other methods are used to transfer the forces from the dome to the rectangular walls. Byzantine architects developed the use of pendentives and arches to carry the stresses of the dome along the four supports in the walls of a rectangular solid. Other non-circular structures resorted to using flying buttresses and other techniques to make sure the walls supported the dome.

pendentives

The record thus far for the largest dome is held by England's *Millennium Dome*[3], constructed to hold an exhibit that ushered in the 21st century. It is twice as large as the previous record holder, the *Georgia Dome* in Atlanta, GA. However, the record for the largest masonry dome is still held by the *Cathedral of Santa Maria del Fiore*[4] in Florence, Italy, with its famous *Brunelleschi Dome*. Filippo Brunelleschi (1377-1446) not only designed the dome but invented machines specifically for its construction. In 1418, a competition was announced for models for the dome's construction. The model was supposed to demonstrate the capping-off of the cathedral along with the methods and devices to be used. Various solutions were presented—from elaborate hanging chains to redirecting forces of the dome to filling the structure with earth and then forming a mound that would allow the dome to be built over the mound, finally removing the earth after the

dome was completed. But the most ingenious solution was that proposed by Brunelleschi. To the amazement and disbelief of the judges and other architects, Brunelleschi contended the dome could be built without any centering, the framework would be used to support the masonry as the construction of the dome rose.

Brunelleschi's design consisted of not one, but two nestled domes enclosing on opposite sides a skeletal frame composed of circular vertical arches (made of 24 semi-arches meeting at a ring on the top) and 9 horizontal rings running between

Photos showing access between the double dome construction.

between the 24 semi-arches made from stone and held together by metal clamps along with tie rings made of oak joined by metal connectors. The outer octagonally shaped dome was designed to protect the inner circular dome from the elements. The first 46 feet of the two domes' shells were made of stone. As the angle of curvature increased in the upper part, it eventually produced a tangent with an angle of 60° to the hemisphere and the masons switched to bricks. To lay the bricks and stones of the domes' shells, Brunelleschi had the

Some of the original and reconstructed tools that were used to built Brunelleschi's dome are currently on display at the Cathedral of Santa Maria

masons use a herringbone pattern. Both friction and other forces on the bricks in the herringbone pattern interlocked one to the other, adding to the stability of the dome. Many feel that Brunelleschi's study of ancient architecture and the Roman Pantheon helped him devise the elements of his ingenious ideas and design.

The spaces between the two domes allowed the masons to work between the domes as the two shells were being constructed from their bases upward. Even today, the same stairwell is accessible, allowing one to witness the amazing dome skeleton and brickwork.

The dome would have been impossible to undertake without Brunelleschi's inventions of machines necessary to raise the building materials to the lofty levels of the dome. The stones weighed between 100-1700 pounds each. After Brunelleschi's death, his incredible ox-driven hoist was studied and copied by architects and artists including Leonardo da Vinci. In addition, Michelangelo carefully studied Brunelleschi's design and devices before undertaking St. Peter's Church in Rome.

Just as the mathematics and physics of the domes held the keys to the properties of domes, spheres and arches, they continue to unlock mysteries of new shapes and forms making it possible for these to be realized in architectural structures.

---

[1]The term dome comes from the Greek word doma, meaning roof, which was adopted into the Latin language.

[2]The Roman arch is based on the shape of a circle. Over the centuries, architectural arches have varied from the circle to the ellipse to the pointed arch. Each shape directs the forces of the weight of the arch differently.

[3]It encompasses 861,000 square feet (80,000 square meters). Its diameter is 1,050 square feet (320 sq. m) and height is 165 feet (50m). Its dome roof is coated in teflon and 2600 cables suspended from 12 steel masts support it.

[4]Construction of the cathedral was started in 1296 and was finished in the 15th century when it was topped-off with Brunelleschi's dome. For those wanting to know more about its history, a fascinating and readable account on its construction can be found in *Brunelleschi's Dome* by Ross King.

are plants counting?

Consider cellular automata, distributed computing, non-linear dynamic systems — Can plants in some way be connected to these areas of mathematics? That's one of the many problems that mathematicians and biologists are exploring in the fast emerging field of biomathematics. *Biomathematics* is a field that uses mathematical techniques and models to explain biological phenomena. Biomathematics is finding niches in such diverse areas as physiology, ecology and even toxicology.

Looking around at plants in our gardens, it's difficult to imagine that calculations may be taking place. Today scientists are speculating and studying various aspects of the dynamics of plant life. At Utah State University in Logan, physicist David Peak, biologist Keith A. Mott, and graduate students Jevin D. West and Susanna M. Messinger have been studying the possibility of plant computations. Specifically they have focused their work on stomata found on leaves and other parts of plants. *Stomata* are pores which allow carbon dioxide to enter and at the same time let water exit. A plant somehow gauges its environmental conditions so that it gets enough carbon dioxide for photosynthesis while at the same time trying not to dehydrate itself. How does a plant maintain this delicate balance? The

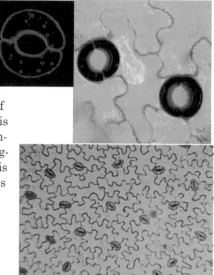

Photos 1 and 2 were taken using a confocal microscope. Photo 1 shows a pair of guard cells of a broad bean (Vicia faba) which is about 50 micrometers long. Guard cells occur in pairs and encircle a stoma (plural stomata). By swelling up they prevent the passage of water and CO2 through the stoma. Stomata occur on the surface of a leaf at approximately 200 per square millimeter. At a lower magnification photo 2 shows two guard cell pairs. In photos 2 and 3 we see jigsaw like puzzle pieces, these are epidermal cells of the leaf.

team at Utah State University has been experimenting with certain plants and has discovered that possibly the process a plant undergoes can be both described and explained by considering it as an optimization problem connected to the workings of cellular automata and distributed computing. An *optimization problem* involves finding the best combination of elements for a problem in which there are various constraints. For example, in complex dynamic systems of plants it can mean maintaining a delicate balance between the right amount of sunlight, water and nutrients in order to prevent a chaotic and devastating event. Since plants don't have computers with central processors to coordinate their functions, scientists are looking to distributed computing to possibly explain how a plant maintains optimum conditions. In *distributed computing* a task is broken down into many smaller tasks, each of which is handled by a single computer unit. Each unit sends its results to a central computer to analyze and solve problems. For plants it is felt these units are its stomata. How is all this individual information in plants processed without a central processing unit? Here is where *cellular automata* enter the picture. Instead of each pore operating independently, these scientists have observed that pores seem to synchronize and work together on various sections of a leaf creating patches where the pores are open and other patches where they are closed, thereby maintaining a healthy system. The patch patterns have been found to be analogous to patterns which evolve on a computer monitor when certain cellular automata are generated. But unlike cellular automata where a central unit is needed to begin, read and process the results, in the case of plants, stomata are not connected by any neuronal tissues which can transmit signals.

Distributed computing—A task is divided into smaller problems which are distributed among a system of personal computers, which gathers the results and produces the solution. This sort of division of labor works best with problems involving vast quantities of numbers and data crunching.

Cellular automaton is composed of a grid of discrete cells which are occupied by discrete values that follow a specified rule(s) at a specific interval of time. For example, on a computer monitor the discrete cells can be represneted by pixels, their discrete values are the binary digits, 0 and 1, which convert to black or white pixels by following the rule(s). For pixels, the rule may be a computer algorithmn which dictates under what conditions a black pixel changes to a white one or vice versa. For pixels the rule(s) may be contingent on the values or color of neighboring cells. A computer model cellular automaton is 2-dimensional, but cellular automaton may be any dimension. In essence, a cellular automaton starts out with an arrangement of discrete cells and as time proceeds the arrangement changes according to its rule(s).

Emergence refers to how unpredictable patterns emerge from interactions between parts which seem independent of one another. Emergence occurs in living and nonliving phenomena. In weather systems emergence evolves from the mingling of carbon dioxide, water, and other molecules. In plants or any other living organism the interactions between cells direct that organism's behavior.

How this dynamic cooperation of stomata is carried out in plants is still a mystery. But work in biomathematics is pointing to *emergent distributed computation*, a form of computing that arises from distributed computing. Stomata are self-organizing and self-regulating entities. Each work as a cellular automaton (guided by some preset rules) with the surrounding stomata to balance the carbon dioxide and water. In the experiments by Utah State University scientists, biologist Keith Mott points out that *"In the case of leaves, stomata are simultaneously the sensors of external information, the processing units that calculate how gas exchange regulation should occur, and the mechanisms for executing the regulation. Thus, in those plants that solve the dilemma of optimal gas exchange, evolution may have found an elegantly parsimonious computational technique in which input, output, and processing are all accomplished by using the same hardware."[1]*

David Peak, physicist of the team, explains that although *"[Plants] are simpler than DNA and they're simpler than brains and if we can demonstrate that plants ... are performing computation, we will have demonstrated for the first time the reality that there is computation in living systems."[2]*

The next time you bend down to smell a flower's bouquet, take time to consider that perhaps precise computations may be taking place within the plant.

The Utah State University team from left to right: David Peak, Susanna Messinger, Jevin West, Keith Mott.

---

[1]Published online January 19, 2004, 10.1073/pnas.0307811100 PNAS | January 27, 2004 | vol. 101 | no. 4 | 918-922 Articles by Mott, K. A. *Computer Sciences / Plant Biology* Evidence for complex, collective dynamics and emergent, distributed computation in plants.

[2]Plants may compute, scientists say by Anita Hansen, The Utah Statesman, (University of Logan) Campus News issue: 3/3/2004.

are you up to the second?

Most of us are familiar with leap year,[1] but few people are familiar with leap seconds.

Scientists have discovered that the Earth's rotational speed fluctuates as the Earth travels through space. They have shown that the tidal action created by the Moon slows the rotational speed of the Earth, adding anywhere from 0.0015 to 0.0020 of a second to the length of a 24 hour day each day per century. It certainly doesn't seem like a big deal to lose a second every now and then.[2] *Why is a second so important since we don't have any way to control the Earth's astronomical clock and add a second back to synchronize its rotational speed?* In today's world, a clock's precision is very important. Important for navigation. Important for communication. Important for satellites placed in orbit to transmit signals such as GPS information or communication signals to points all over the Earth. Time accuracy is also very important to *geodesy*, the science that deals with the shape, size and curvature of the Earth. Geodesy, in essence, surveys the Earth, explaining how to measure great distances and locate points on the Earth's surface by using *triangulation*. Precision timing is invaluable to astronomers and space travel. Accurate time is crucial to air traffic control systems and global computer activi-

ties such as international finances. Synchronized clocks are needed for high speed communication systems and power grid distributions. Consider what would happen if our radio and television programs were not precisely coordinated.

> ***Triangulation*** is based on the properties and theorems of the triangle as described in Euclidean geometry. For example, if only one side and two angles of a triangle are known its other measurements can be calculated. The ancient seafarers and travelers used triangulation to navigate even as GPS (global positioning system) works to pinpoint locations and whereabouts of objects.

So what has been done? Today our clocks' time, the time on our cell phones, the time on our VCRs are set from signals according to atomic time, which is determined by a network of atomic clocks that are stationed at various worldwide locales in environmentally controlled chambers. The TAI (Temps Atomique International), International Atomic Time, is the average of the times of these atomic clocks. This averaging is carried out by BIPM (Bureau International des Poids et Measures) which is responsible for coordinating the process. These atomic clocks keep the precise time by relying on the UTC second or atomic second—defined as 9,192,631,770 oscillations of a cesium 133 atom.

Astronomical clocks, on the other hand, are based on the Earth's daily rotation. In 1972 the UTC (Coordinated Universal Time) scale was adopted internationally. The astronomical second was defined as 1/86,400 of a day as determined by the average solar day for the year 1900 and designated as the UT1 second.

Occasionally, atomic clock time(UTC) and astronomical clock time (UT1) differ ever so slightly. Yet this difference can have profound effects. Since there is no way to speed up or slow down astronomical clock time, every so often a second is added to atomic clock time thus creating what is now called the *leap second*. The IERS(International Earth Rotation Service) in Paris is responsible for keeping the UTC time to within ±0.9 second of the UT1 time. To carry out this task, the IERS announces when a leap second will be inserted and informs the various international agencies. The international agencies have agreed that a leap second

Whether it's this 18th century spring pendulum clock or the oscillations of an atom's electrons in an atomic clock, seconds rhythmically tick away.

can only be added to the end of December or June or if needed to the end of March or September. This procedure has been in effect since 1972 when the first leap second was announced for June 30, 1972. Ever since then, 22 leap seconds have been added to UTC over 27 years; and thus far, these have only been added in June and/or December. In theory, a leap second can be negative; but to date, a leap second has never needed a negative adjustment.

It's amazing to learn that the Earth has been on time for the past five years with no need to add leap seconds. Scientists have no definite explanation for this phenomenon. They have shown that the tides generated by the Moon create friction that slows the Earth's rotational speed. Other factors believed to affect the Earth's rotational speed include large earthquakes, yearly melting in the spring of snow and ice in the northern hemisphere, changes in large bodies of water in dams and reservoirs, global warming, changes in climate and in the Earth's core.

Mathematics has played an invaluable role in measuring, calculating and explaining time, the effects of the Earth's rotational speed and leap seconds. Incidentally, even though 2004 was a leap year, there was no leap second for June 2004, but one was added on December 31, 2005. If you really want to stay on time, call 202-762-1401 in the Washington, D.C. area for a time of day[2] announcement. A private commercial service is also available at 900-410-TIME.

A cesium fountain atomic clock developed at the NIST laboratories in Boulder, Colorado. Photo courtesy of NIST.

---

[1]Since it takes the Earth 365 days 5 hours and 48 minutes and about 46 seconds to complete its solar year, the 365 day year doesn't do the job accurately. In 1582 Pope Gregory XIII revised the Julian calendar that had been in effect since 46 BCE because it gained 1 day every 128 years. To deal with the ever accumulating discrepancy between the date and the Earth's position in its orbit, the Gregorian calendar was set up to have 365.2425 days per year and included a revision in the leap years. A leap year is created every four years when a day is added, but this causes approximately 3 extra days every 400 years. To adjust for this it was decided that any leap year that lands on a century year that was not divisible by 400 be eliminated. These adjustments lessened the discrepancy to 1/4 minute every year, which meant one day every 3,323 years.

[2] There are a number of websites which provide atomic times. Check out http://www.worldtimeserver.com and http://www.time.gov.

Sometime between late 2002 and early 2003, Alex Matter found thirty-two art works wrapped in brown paper in a storage unit in Long Island. Attached was a note written by his father Herbert Matter, which identified the pieces of art as early experimental paintings by Jackson Pollock.

Were these authentic works of Pollock? Early on, Ellen Landau, Case Western Reserve University art historian, authenticated them as Pollock's work. In 2006 mathematics was brought into the analysis. Later on in 2007, material analysts were also called in to examine these works. An analysis done at Harvard University found in three of the works pigments and binders which were not commercially available during Pollock's lifetime. Could the works have become contaminated when they were restored after their discovery? Another analysis was done by James Martin with Orion Analytical. As yet his findings have not been allowed to be made public. Will these findings cast doubts as to whether any of the 32 works were by Pollock?

How did mathematics figure into this controversy? In June of 1999 *Nature* magazine published *Fractal Analysis of Pollock's Drip Paintings,* a study done by physicist Richard Taylor.

Upon studying 17 of 180 Jackson Pollock drip-splatter works, Taylor claimed his work demonstrates that Pollock's drip-splatter paintings have their own fractal signature, and he contends they conform to a specific statistical fractal dimension unique to Pollock's work. An important property of fractals is *self-similarity*— its whole shape is evident in any of its parts, that is, its shape continually recurs as one zooms in or out of a fractal. This is easily identified in geometric fractals, such as the *Sierpinski triangle* in which a shape's pattern is continually repeated. In some fractals self-similarity is not immediately apparent. In these fractals mathematicians look at statistical repeti-

For natural phenomena such as this cloud formation, mathematicians must use statistics to determine its self-similarity and fractal dimension.

tion to quantify a fractal's dimension. Fractals can be used to describe complex shapes which defy Euclidean geometric figures. They can also be used to recreate various growth patterns and shapes such as those found in cauliflower, ginger, clouds, mountain ranges, coastlines; that is patterns found in natural phenomena. Complex fractals are categorized by their fractional dimensions, which range between one and two. If they were 1-dimensional, the length between any two points would be finite, but a fractal segment is infinite because of its self-similarity property it is in essence continually replicating itself between any two of its points. Consequently, the closer a fractal's dimension is to 1 the less complicated its shape, the closer to 2 the more complex its shape.

The Sierpinski triangle is not a random fractal and its fractal dimension can be easily cal-culated by looking at a portion of it and notic-ing that the red part is repeated 9 times in the portion shown and that it is 1/4 the size of the entire portion. These number are then applied to a formula that gives its fractal dimension as ≈1.5849.

In 2006 the Pollock-Krasner Foundation, which is in charge of the Pollock estate, wanting to determine the authenticity of the 32 paintings, turned to Taylor and his scientific method of analysis, since Taylor claimed he could identify an authentic Pollock work by using fractal geometry. Applying techniques from his 1999 study to 6 of the 32 paintings in question, he concluded that they did not possess the fractal characteristics that he claimed were unique to Pollock. His findings were published in both *Nature* magazine and *The New York Times*.

Earlier in 2004 graduate astrophysicist student Katherine Jones-Smith looked closely at Taylor's work, *Fractal Expressionism*, and realized she could easily produce sketches that were "fractal" according to Taylor's criteria. In 2006, when she learned that Taylor had been commissioned by the Pollock-Krasner Foundation to analyze some of the Matter paintings with his fractal analysis, she knew it was important to share her findings. She and Case physics professor, Harsh Mathur, coauthored *Fractal Analysis: Revisiting Pollock's drip paintings,* which appeared in the November 2006 *Nature* magazine. Here they proved Taylor's method was flawed by creating a simple drawing which passed Taylor's criteria for a Pollock work. Jones-Smith has shown that Taylor's research was deficient on several levels. She explains that Taylor claimed Pollock's *drip paintings* were created using a certain chaotic motion that left fractals trails, yet as Jones-Smith points out, other chaotic motions which are known *not* to be fractal (e.g. space filling curves) appear fractal when observed under the limited range of magnification that Taylor used. She explains that since Pollock's work does not have visual self-similarity like those apparent in the Mandelbrot set or the Sierpinski triangle, he had to use statistics to derive Pollock's fractal dimension. Yet, she explains that the degree of magnification Taylor used was not great enough to determine the fractal dimension and to identify true self-similarity. Jones-Smith points out that *"Virtually anything appears 'fractal' over 2 orders of mag-*

nitude if your definition of fractals is simply the demonstration of a non-integer box counting dimension....over a limited range ...[this] is not enough to distinguish fractals from things which really have no

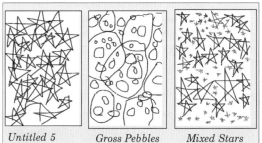

Untitled 5   *Gross Pebbles*   *Mixed Stars*

Images created by Katherine Jones-Smith which illustrate the same fractal signature of Jackson Pollock as defined by the criteria presented by Richard Taylor. *Untitled 5* appeared in *Revisiting Pollock's drip paintings by* Katherine Jones-Smith and Harsh Mathur. *Nature* Nov. 30,2006.

business being called 'fractal'." Furthermore, Jones-Smith and Mathur point out that *"In short, we have shown that the "fractal" characteristics Taylor espoused as being unique to Pollock can be created at will and are as equally present in such crude sketches as Untitled 5 as they are in any of the Pollock paintings Taylor studied; also we exposed severe mathematical inconsistencies[1] in the assumptions Taylor used to formulate his hypothesis that Pollock's paintings are fractal."* She also emphasizes *"if one was to use only fractal analysis, one would conclude that all of the three sketches [above] are authentic Pollocks. Obviously they are not; which is why we believe fractal analysis should not be included in the debate*

surrounding the authenticity of the paintings in the Matter cache." In addition, in the most recent paper (October 2007) by Jones-Smith and Mathur, *Drip Paintings and Fractal Analysis,* examples of two commissioned drip painting forgeries are shown that pass as authentic Pollock's using Taylor's analysis. Essentially, Jones-Smith and Mathur have shown that Taylor's fractal analysis cannot be used as an art authentication tool.

On March 4, 2008 Taylor et al wrote a reply to the challenges to his fractal work. Jones-Smith, Mathur and Krauss immediately responded and addressed each of Taylor's rebuttals. Now it is up to Taylor to respond with specifics to their conclusions.

The 32 disputed paintings, along with other works by Pollock, were shown at the special exhibit *Pollock Matters* at the McMullen Museum on Boston College from September 1 through December 9, 2007. Ellen Landau was the curator. The show displayed the pieces of the Pollock-authenticity puzzle with evidence from both sides of the controversy discussed. The exhibit then went to the Emerson Museum of Art in Syracuse, New York from December 21, 2007 until March 20, 2008.

Thus far *Whodunit* remains unanswered.

---

[1]Among the mathematical inconsistencies found were those dealing with chaotic motion, the superposition of two fractals, and chi not being an intrinsic property of the image.

math & your ring tones

Today math is even invading the ring of your cell phones. The colorful bars shown here are visual

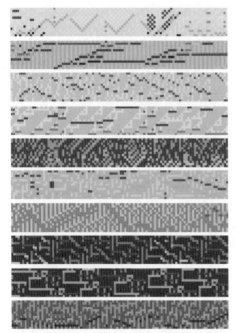

Diagrams of notes selected from cellular automata—
http://tones.wolfram.com/

representations of music created by mathematics. It's difficult to realize, let alone believe, that in the world of cellular automata music can be found and that these digital entities are produced by simply using pixels and simple rules. Yet music and much more have been discovered in the pixels of the most basic one-dimensional cellular automaton[1].

In the 1980s Stephen Wolfram devoted himself to exploring and experimenting with the artificial worlds of cellular automata. In the process Wolfram uncovered connections between the artificial universe of cellular automata and our universe. He developed a mathematical system using these digital creatures, and found connections between their patterns and how they relate to things in our universe, particularly to patterns in nature. Wolfram called his work *A New Kind of Science,* and he explained it in the over 1000 pages of his book by the same name. In conjunction he also conceived *Mathematica,* a very sophisticated mathematics

Stephen Wolfram.

program which allows you to solve and explore not only simple mathematical problems but complex problems and concepts as well.

Now you can experiment, explore and enjoy a small segment of *A New Kind of Science* by creating your own personal music. Log on to *http://tones.wolfram.com/* and compose music using cellular automata, which can then be downloaded to your cell phone. The website is an exciting place to explore music and mathematics. You select the musical styles, instruments, the tempo, percussion, etc.. Here cellular automata and music theory are integrated by *Mathematica* algorithms converting them to music which you hear and musical scores you see produced in the patterns of these colorful bars. As the website points out *"Each program in effect defines a virtual world, with its own special story—and WolframTones captures it as a musical composition."* You'll want to spend a lot of time exploring this website and all its links, intricacies and information. Whether you just want to compose and download your ring tone to your cell phone or explore the musical world formed by a cellular automaton and its mathematics, it's all there. The possibilities are remarkable. It's an experience you'll not want to pass up.

The idea of **cellular automata** originated with Konrad Zuse and Stanislaw Ulman in the 1940s. Using the concept of cellular automata John von Neumann showed how a computer could be programmed to self-reproduce. Then in the 1970s, interest in cellular automata was renewed with the premiere of John Conways' *the game of life*, and interest was boosted further in 1980s by the work of Stephen Wolfram.

What do pixels on a computer screen have to do with music? Look at this string of numbers …00100000… and apply this simple rule: after each moment of time a digit is replaced by the sum of itself and the neighbor on its right. This produces …0110000…. Each string of digits could be thought of as a horizontal line of pixels on a screen with 0 being a white pixel and 1 a black. This is a numerical form of a 1-dimensional cellular automaton. Now look at just the string of 3 pixel cells. The possibilities are:

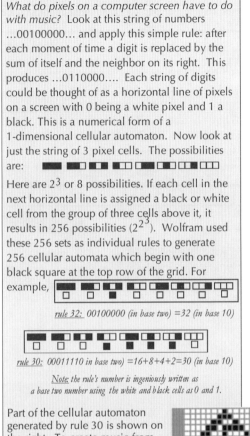

Here are 2^3 or 8 possibilities. If each cell in the next horizontal line is assigned a black or white cell from the group of three cells above it, it results in 256 possibilities (2^{2^3}). Wolfram used these 256 sets as individual rules to generate 256 cellular automata which begin with one black square at the top row of the grid. For example,

rule 32: 00100000 (in base two) =32 (in base 10)

rule 30: 00011110 in base two) =16+8+4+2=30 (in base 10)

Note: the rule's number is ingeniously written as a base two number using the white and black cells as 0 and 1.

Part of the cellular automaton generated by rule 30 is shown on the right. To create music from this, a swath (the middle section) is taken and laid on its side as shown below.

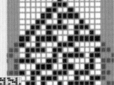

Time runs horizontally across this swath, while pitch is represented by vertical columns. Images are from "A New Kind of Science" by Stephen Wolfram—http://www.wolframscience.com/

Music is composed by using *Mathematica* algorithms to map the pitch to the notes of a selected scale.

For more in-depth information on how Wolfram tones work see the website at http://tones.wolfram.com/.

surfing with mathematics

We've heard of skate boarding parks and ice skating rinks, now surfparks are emerging worldwide. What's behind it all? ... mathematics and Kerry Black and his team. Black combined his knowledge of

the science of wave theory with the zeal of a surfing enthusiast, and came up with the idea and a method to replicate time and again the perfect surfing wave. Having extensively studied mathematics and with degrees in physics, geophysics, and oceanography, he and his team put their expertise to work and designed a special wave system. This system uses a specially designed wedge pool that is controlled and powered by computer software which also controls the pool's adjustable floor.

Black points out that *"you have to have an intimate knowledge of surfing and wave theory that allows you to forecast how tubey or how fast a wave will be."*[1] While surfing primo destinations around the globe, he researched, studied and recorded wave dynamics and oceanography of varied regions which helped him unravel the mystery of the perfect surfing wave. Using mathematical modeling and computer simulations in conjunction with his research on reefs, sand movement, breaking intensity of waves, peel angle (the angle between the trail of broken white-water and the unbroken crest), Black and his team created a viable working surfpark in various parts of the world. Black explains that the surfpark's wave action can be changed to accommodate beginner to seasoned surfers with an infinite possibility of wave shapes.

Artificial surfparks[1] are not intended to replace real-life surfing, but instead provide an alternative, and perhaps will help make it possible for surfing to become an Olympic event.

---

[1]To view a surfpark and its components online go to www.surfparks.com

tackling murphy's law with math

"*If anything can go wrong, it will go wrong.*" is one way to express the famous adage known by such names as *Murphy's Law, Finagle's Law, Sod's Law.* Some people consider it a myth others take it seriously. British mathematician Philip Obadya, working with colleagues David Lewis and Keylan Leyser, came up with a formula that statistically calculates the likelihood of this law. Working with a sample of over 1000 people, Obadya's equation is:

was determined by polling over 1000 people.

F stands for how *frequently* you perform the task, and is also assigned a value from 1 to 9.

The Rating of Sod's Law, R_{SL}, ends up ranging between 0 and 8.6, where the higher number warms you that it's likely something may will happen.

Obadya points out in *Null Hypothesis, The Journal of Unlikely Science* that *"The lesson from this is*

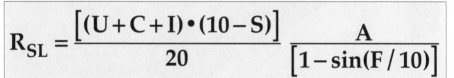

$$R_{SL} = \frac{[(U+C+I)\bullet(10-S)]}{20} \frac{A}{[1-\sin(F/10)]}$$

To figure out the likelihood of the law occurring, you assign value to the variables in the formula as follows:

U stands for the *urgency* of a task and is given a value on a scale of 1 and 9, with 9 meaning most urgent.

Similarly for **C** stands for the *complexity* of a task and is assigned a value between 1 and 9.

S represents how *skilled you are* at performing the task, and also is assigned a value between 1 and 9.

A stands for *aggravation*, and is a constant. Its value is 0.7, which

that, to cut the seemingly unbeatable Sod's Law Gremlins down to size you need to change one of the elements in the equation. .... There is, of course, a Sod's Law element to using the equation as well. So beware. If you judge your ratings wrongly you might become too optimistic, allowing calamity to strike. Furthermore, knowing a priori the RSL value of a particular task may well lead to over-confidence, producing a positive feedback mechanism by which the Sod's Law rating increases still further."

the sierpinski triangle find its way into the grand egyptian museum

Sometime in 2009 the Grand Egyptian Museum (GEM) is scheduled to open its doors. The first most striking feature of GEM's design is its facade. Stretching over 1800 feet, the Sierpinski triangle will adorn the museum's front. The triangular faces of the Great Pyramids will thus be echoed time and again by this ever evolving geometric fractal. This facade speaks of mathematical infinite self-replicating triangles. It will not only greet some 15,000 visitors daily, but be a beacon at the junction of Egypt's dry desert and the Nile's fertile valley. In addition, the landscape's dunes will be captured by the roof's structural folds and two opposite edges of the roof will align themselves with vertices of both the Cheops Pyramid (the largest) and with the smallest of the three pyramids, the Khefre. The translucent facade allows the Sun's rays to flood light into the museum. At night the facade will be illuminated and connect with lighting of the Great Pyramids so that the Sierpinski triangular facade will appear to glow. This fractal triangle of triangles is named after Polish mathematician Waclaw Sierpinski who

A rendition of how the front facade of GEM's entire translucent stone wall will appear. Design: heneghan.peng.architects | Arup | Buro Happold JV. Coutesy and © by heneghan.peng.architects.

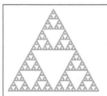

Early stages in the formation of the Sierpinski triangle

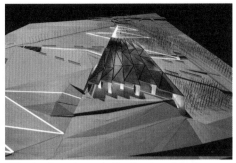

The view from the south (the direction of the Pyramids) showing the structuring of the project according to the lines of geometry from the site to the pyramids. Design: heneghan.peng. architects | Arup | Buro Happold JV. Courtesy and © by heneghan.peng.architects.

mathematically described it in 1915. He was one of several mathematicians who were intrigued by what were then considered mathematical anomalies, and are now known as fractals.

The term *fractals* was coined by Benoit Mandelbrot, who first discovered a link between fractals and a natural phenomenon. *What is a fractal?* A fractal is made by using an object which may be a geometric object, such as a square, or another type of mathematical object such as a number. A set of rules called an *algorithm* or a formula is applied to the object which produces a new object. Then the rule is applied again to the new object, and another object is made. The rule is applied again, and again and again.... The process of reapplying the rule or algorithm is called an *iteration*. The result may be a geometric fractal such as the *Sierpinski triangle* that exhibits symmetry, or it may result in a random fractal, such as the

GEM is being built about one mile from the Great Pyramids of Giza, and will symbolize a bridge which connects Egypt's ancient majestic pyramids and modern Cairo. Egypt's original museum in Cairo, *The Egyptian Museum*, was initially designed as the home for some 10,000 artifacts when it was built over a century ago. With ongoing archeological discoveries, Egypt's treasures have multiplied over fifteenfold. Today, *The Egyptian Museum* cannot do justice let alone adequately house these artifacts. Hence, the Egyptian government announced an international architectural design competition for GEM. In 2002 a site of 117 feddans[1] was selected for the new museum on the first plateau between the Great Pyramids of Giza and Cairo. With over 1557 designs from 83 countries, the competition turned out to be the largest ever global architectural design competition. The award was announced in June of 2003, and its winner was Heneghan. Peng. Architects[2]. One of the most difficult tasks in GEM's design, as architect Roisin Heneghan explains, was *"How do you build a major museum in its own right beside the greatest structures that have ever been built? If it's too dominant, you start to take away from the environment in which the pyramids were set."* To accomplish this she explained *"There's a 150-foot difference in levels at the site, so we embedded the museum between them, creating a new cliff face. Coming out of Cairo, you never see the building at the same level as the pyramids. Plus, from inside the museum itself you can see all three pyramids.*"[3] The cost of GEM is estimated between $350-500 million dollars and is set for completion in 2009.

The GEM's structure will cover over 409,000 square feet of exhibit halls, which will not only function as a repository for many of Egypt's over 100,000 ancient treasures, but will also act as a research, conservation and education center. In addition, it is designed to give visitors both real and virtual insights and experiences into the times and life of ancient Egypt. The visitor will be swept up by its permanent exhibits and also be virtually transported to the archeological discoveries. The GEM will act as a bridge to the past from the present—a junction from modern Cairo to the great Pyramids. Even the Grand Staircase that ascends through the museum will be both a focal orientation point and a portal through time leading the visitor from the lobby to special exhibitions, conservation workshops and temporary exhibits and eventually to the permanent galleries. Here collections will be presented in chronological order as one ascends until reaching the top where the visitor will encounter breathtaking views of the Pyramids of Giza, thereby making these pyramids part of the museum's permanent exhibit.

In preparation for the work on GEM, the 83 ton statue of Ramses II was relocated in August 2005 from its

Mandelbrot set or non-linear fractals which possess what is called statistical symmetry. Whichever type of fractal you get, it will exhibit the property called *self-similarity*, meaning whichever part of the fractal you zero in on or magnify will resemble the original part.

The Sierpinski triangle can be described mathematically in a number of different ways. A simple way is to start with any shaped triangle. The algorithm is:

• Connect the three midpoints of the triangle's three sides. This results in the formation of four smaller congruent triangles in the original triangle's interior.

• Then remove the center triangle.

• Now reapply the algorithm to each of the 3 small triangles that surround the triangular hole in the center of the original triangle.

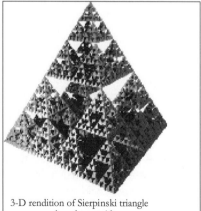

3-D rendition of Sierpinski triangle as a square based pyramid.

• Continually iterating the algorithm to the remaining triangles results in the Sierpinski triangle.

The same process can be applied to a square based pyramid, such as the Great Pyramid of Giza. The result is a 3-dimensional version of the Sierpinski triangle, shown on page 74.

It is fascinating to explore some of the many different methods thus far developed to generate the Sierpinski triangle. These include using Chaos game and Pascal's triangle. The Chaos game method starts with any triangle. Half the distance between a point chosen inside the triangle and one of the triangle's vertices (selected at random) is drawn as a new point. When this algorithm is continually repeated, the Siepinski triangle emerges.

Cairo location at the main train station to the Giza plateau. Here the over 3200 year old statue is being cleaned and prepared for its installation within the main triangular gated entrance.

The museum will be an amazing complex which weaves exterior space with its parks and piazza transitioning them effortlessly into its entrance and lobby. There is no doubt that the Grand Egyptian Museum will awe its visitors with its beauty, setting,

---

[1] The term feddan comes from the Arabic word meaning *'a yoke of oxen'* and refers to the area of earth that can be tilled by oxen in a certain time. In Egypt the feddan is the only non-metric unit that remains in use following the switch to the metric system.

1 feddan = 24 kirat = 4200 square meters = 1.038 acres

[2] Heneghan Peng Architects was established by Roisin Heneghan and Shih-Fu Peng in New York in 1999, and relocated to Dublin in 2001.

The architects involved in GEM's design are: *Team Leader:* Shih-Fu Peng, Heneghan.Peng.Architects and *Team Members:* Roisin Heneghan, Edel Tobin, Alicia Gomis-Perez, Arup, Buro Happold, Bartenbach L'chtlabor GmbH.

[3] From FAST COMPANY.COM 5th Annuta Fast, number 15 *Monumentalists.*

the misplaced manuscripts fiasco

There are two infamous goofs by mathematician Augustin Cauchy that probably delayed the progress of mathematics and had detrimental effects on the lives of two outstanding mathematicians. In both instances Cauchy was entrusted with the powerful and ground breaking work of young mathematicians Evariste Galois and Niels

ty for Galois' talent and work to be recognized. The rest of Galois' short life was a set of tragic circumstances. Plus he never received any recognition for his extraordinary mathematical work during his life.

In 1828 Niels Abel(1802-1829) also gave his work on transcendental functions to Cauchy. Again Cauchy misplaced it, even as he had misplaced the work of Galois[1].

Had Cauchy not been so careless might mathematics in these areas have progressed more quickly? Would Galois' and Abel's lives have had different turning points? Would they have gone on to discover more amazing mathematical ideas?

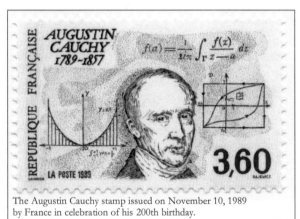

The Augustin Cauchy stamp issued on November 10, 1989 by France in celebration of his 200th birthday.

Abel.

In 1829 Galois(1811-1832) gave his paper containing his fundamental discoveries on the theory of equations to Cauchy who promised to submit it to the French Academy of Sciences. Not only did Cauchy not present it, claiming he forgot, but then said he had lost the work thereby missing an early opportuni-

---

[1]See pages 135-137 for details on Niels Abel's life.

the inheritance problem

Inheritance and the division of an estate has its difficulties regardless of how thorough one's estate planning may be. This old problem of Arabic origin has appeared in many different versions. Here is the camel version.

Before he died, a merchant willed his 17 camels to his three sons. The camels were to be divided among them as follows: The eldest son was to receive 1/2 of the camels, the middle son 1/3, and the youngest 1/9. The executor was perplexed. Finally, he devised a method for dividing the camels without having to slaughter any. What did he do?

SOLUTION:
The executor added one of his camels to the 17. Naturally the sons did not complain, since any fraction of 18 would be more than of 17 camels. So the eldest son got 1/2, namely 9 camels, the middle son got 1/3 which was 6 camels, and the youngest got 1/9 or 2 camels, =9+6+2=17 camels, and the executor took his camel back.

Can you come up with other sets of unit fractions (that is with numerators 1) and total number of camels to divide among these three heirs with the executor always adding one of his to solve the problem?
For example,
1/2, 1/3, 1/7 with 41 camels.
1/2, 1/3, 1/8 with 23 camels.
1/2, 1/4, 1/8 with 7 camels.
Can you find others?

77

the möbius strip continues to twist our fancy

August Ferdinand Mobius (1790-1868) was a professor of astronomy who wrote popular treatises on principles of astronomy and Halley's Comet as well as textbooks on statics, celestial mechanics, the theory of circle transformation and numerous other mathematical topics. Today, however, he is remembered for his 1858 discovery[1] of an unusual twisted band called the Möbius strip or Möbius band.

What was and is so special about this twisted ring? It illustrates how a 2-sided object, such as a strip of paper which has a front and a back side, can be converted into a ring that is single-sided and single-edged.

• Take a strip of paper and twist one end 180° or a half-twist and join the two ends together. With a pencil, trace a line down its middle. You will trace both sides and return to the starting point; and unlike a normal ring, you never lift the pencil!

• The same thing happens with your finger tip tracing the edge. The Möbius strip's entire edge is traced without lifting your finger. To trace the edges of a regular paper ring, your finger tip must be moved from the edge on the left of the ring to the edge on its right.

At first encounter, the Möbius strip seems like just a mathematical curiosity. But it was the Möbius strip that gave topologists a concrete model of a one-sided object. Sidedness is a topological characteristic called an invariant trait.

The Möbius strip not only entertains but astonishes us with many surprises as one experiments with paper models. Discovering what happens when a paper Möbius strip is cut down the center amazes any Möbius strip novice. The list of

> In addition, in his topological work Möbius proved such things as bands with an odd number of half-twists are one-sided and those with an even number of half-twisted are 2-sided.

Möbius strip experiments and intriguing results go on and on. The following experiment, illustrated in the diagrams at the right, has a romantic touch.

First make one Möbius strip by giving a half-twist clockwise and gluing its ends together. Then make another Möbius strip by giving the strip a half-twist counterclockwise. Attach the two Möbius strips perpendicularly, as shown. Then cut along the dotted lines. Two interlocking hearts are formed.

Many applications have been devised using Möbius strips. Among these are: Richard L. Davis' the nonreactive resister Möbius strip

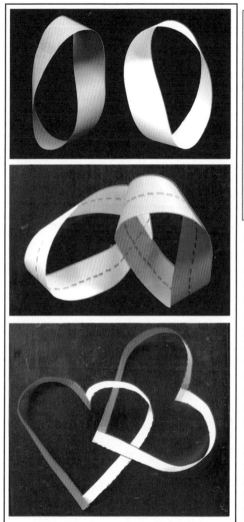

In *topology invariants* are those characteristics which remain unchanged after the object has undergone topological transformations of stretching, shrinking, and scrunching. Invariants are core elements in the study of topology. In addition to sidedness, other topological invariants include inside, outside, and continuity (no gaps or holes occur from the transformation).

•The top frame shows two Möbius strips. One is formed from a counterclockwise half-twist and the other from a clockwise half-twist respectively.

•The middle frame shows the Möbius strips from the top frame glued together perpendicularly.

• The last frame shows two interlocking hearts formed by cutting along the dotted line in the middle frame.

used by the Atomic Energy Commission. The B.F. Goodrich Co. Möbius strip conveyor belt. In addition to being used in stories, novels and films, the Möbius strip appears in works by artist M.C. Escher, in postage stamp art, and graphic art logos such as the symbol for recycling. We even find the Möbius strip design used in earrings, pendants, and scarves.

But what do 21st century mathematics and science have to do with the 19th century Möbius strip? Today the Möbius strip as a scientific concept is being discovered in natural phenomena and used in such fields as chemistry, quantum physics and nanotechnology. Minuscule Möbius strips are being pursued at the atomic level. Atoms, molecules and electrons are being coaxed and sometimes coerced to form molecular Möbius strips. In 1981 David M. Walba of the

University of Colorado synthesized a molecule in the form of a Möbius strip, a double ladder strip made from carbon and oxygen atoms. Since then, numerous other nano-scale Möbius strips have been produced, such as the Möbius cyclacene. Some examples have their molecular framework twisted into Möbius strips while in others only the electrons' wave action mimics the Möbius strip. In nanoworld, Möbius strips can be made into a

The Möbius strip appeared on this Argentinian stamp issued in 2000.

trefoil knot. In 2002 a Japanese team of scientists headed by Satoshi Tanda[2] at Hokkaido University grew Möbius crystal strips of niobium and selenium (NbSe3). They coaxed the crystals to grow around droplets of selenium; and in so doing, they formed rings, some of which twisted 180° thereby forming Möbius strips. These *"...crystal

Möbius was not alone in discovering and studying the Möbius strip. In fact, unpublished works of both August Möbius and Johann Benedict Listing show both men came up with the idea independently in 1858. Listing's work predates that of Möbius by a few months. Listing's source for his topological ideas may have come from Karl Gauss (1777-1855) with whom Listing had studied and corresponded. Listing described the Möbius strip in his 1861 published work, *The census of spatial complexes or generalizations of Euler's theorem on polyhedra*. Although Möbius did not publish his discovery until 1865, it is his name that is associated with the quirky band. See page 111 in *Möbius and his Band* edited by J, Fauvel, R. Flood, and R. Wilson, Oxford University Press, Oxford, 1993.

forms offer a new route to exploring topological effects in quantum mechanics as well as to the construction of new devices."[3]* Where will all these molecular-sized Möbius strips lead? To parts and devices for nano-sized tools? Only time will tell. Regardless, they will guarantee to continue to entertain us and amaze us because the Möbius strip has always been more than a one-sided matter.

---

1 August Möbius discovered his one-sided band/strip while conducting research on geometry theory of polyhedra. He wrote *"while holding side AB fixed, twist the strip through an angle of 180° about the middle line parallel to AB until A'B' is opposite AB, and then bring A'B' into coincidence with AB"*. See page 121 of *Möbius and his Band* edited by J, Fauvel, R. Flood, and R. Wilson, Oxford University Press, Oxford, 1993.

2 Tanda's colleagues include T. Tsuneta, Y. Okajima, K. Inagaki, K. Yamaya and N. Hatakenaka.

3 Nature (2002), 417, 397.

saccheri missed the boat

Sometimes one's thinking is clouded by what one wants to see or hear. Sometimes people are so intent on proving something is so that they do not recognize what's before their eyes. It certainly was the case with Italian mathematics professor Girolami Saccheri (1667-1733). In 1733, the same year he died, he published *Euclides ab omni naevo vindicatus* (Euclide cleared of all flaws). In this work he set out to prove that Euclid's postulate[1] was true by using an indirect proof approach. Using what's come to be known as Saccheri's quadrilateral, he first showed that angles A and B were congruent. Then he assumed that only one of three possibilities could be true about angles A and B: (1) they were both acute angles; (2) they were both right angles; or (3) they were both obtuse angles.

Saccheri's work took him on a long mathematical journey in which he proved many new results, some of which were strange and very unusual, and did not describe or relate to things in the Euclidean geometric world. What he failed to acknowledge was that all three possibilities could hold true in their own very different geometric worlds. He could not bring himself to accept that Euclid's geometry was not the

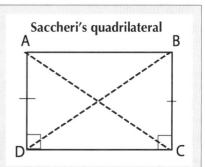

Saccheri's quadrilateral

By drawing in diagonals AC and BD, Saccheri proved angles A and B were congruent by showing ΔACD≅ΔBDC and ΔADB≅ΔBCA.

He then set out to use an indirect proof to show why angles A and B had to be right angles and not either acute or obtuse.

only possible geometry, and, instead, he talked himself into believing this was the contradiction he was seeking. If he had not been so intent on vindicating Euclid, but believed in his own logic and work he would have been credited with discovering a non-Euclidean geometry 100 years earlier than János Bolyai, Nikolai Lobaschesky, and Georg Riemann. Would it have changed the progress of mathematics? Would his findings have been accepted at that time?

---

[1]This refers to Euclid's Fifth (or Parallel) postulate which describes that on the same plane there is one and only one line that can passes through P and is parallel to L.

points of deception

Three examples of anamorphic art. The dancer from *Wonderful Transformations* requires a cylindrical mirror to transform its drawing. The *alien who crashed* and the *boy angler* are a few of the whimsical 3-D drawings by Greg Brown which appear on exterior walls of public buildings throughout the city of Palo Alto, CA.

Anamorphic art transforms reality. It all depends on how you look at it. Some of this art relies on your point of perspective. Other works rely on the means by which you view them. Regardless, anamorphic art can be confusing, intriguing and surprising all at the same time.

Anamorphic art first originated during the Renaissance when artists explored and incorporated the concept of perspective.[1] In addition, at this time mathematics of projective geometry also entered the scene. Initially artists transformed a 3D scene using their eye and its position as the focal point for capturing the image on canvas. Some artists went further to precisely render the exactness of their scene by using such tools as the perspectoscope, the camera obscura and the device below shown in Albrecht Dürer's 1525 engraving.

Albrecht Dürer's 1525 *An artist drawing a nude.* The original engraving was altered to show a non-vertical superimposed screen inserted.

This drawing of a child's face by Leonardo da Vinci is one of the

earliest examples of this type of anamorphic art. In order to view his sketch as an undistorted face, the observer must view it at an oblique angle or position the drawing so that the paper's edge is almost vertical to the viewer's nose.

Over the years various other types of anamorphic art evolved. Some transformed the art on a square grid onto trapezoidal, rectangular, triangular, and even circular grids. Other anamorphic art relied on reflections using varied shaped mirrors, such as cylinders, cones or pyramids to unmorph the drawing and view it as it was originally seen or even what it actually was. The use of mirrors in anamorphic art is believed to have originated in China around 500 BCE where the properties of concave mirrors were first explored.

To create *distorted* art pieces takes patience along with mathematics and physics. Among the areas used are:

• *Topology* works at physically changing the shape of an object by stretching, squashing and pulling it, as seen in how this square was transformed into an arch-like plane.

The square is transformed into an arch.

• *Translating and mapping* moves one set of points to another location by regraphing on a modified grid, as this 3x3 grid illustrates. Another type of mapping transforms an object mapped on a square grid to a circular grid.

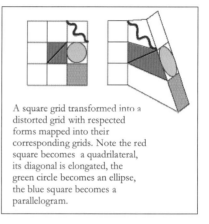

A square grid transformed into a distorted grid with respected forms mapped into their corresponding grids. Note the red square becomes a quadrilateral, its diagonal is elongated, the green circle becomes an ellipse, the blue square becomes a parallelogram.

• In *projective geometry,* one set of points is projected from one plane to another while working with different points of

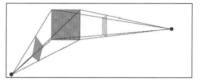

projection(vanishing points) and parallel converging lines, as shown above.

Physics explains how light rays travel in straight lines, and how the eye receives these rays as they all converge to the eye. Think of an artist's canvas as intersecting these converging lines of sight. Think of the image as being preserved and imprinted on the canvas by the artist. The image can be distorted by changing the vision lines. Nevertheless the distorted image is intact and its actual view can be recaptured if the viewer changes his/her perspective point even as with the drawing of a child's by da Vinci. For example, depending on

Various views of a square.

your vantage point a square can appear to be a rectangle, a rhombus, or a parallelogram, but just by moving the position the image can once more become the square. The physics of light and reflection explain how a traditional mirror bends light rays. The incoming ray's angle, called the *angle of incidence,* is reflected at exactly the same angle, called the *angle of reflection.* On the other hand, if the mirror is instead a cylindrical, conical or pyramidal

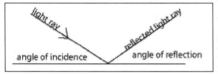

angle of incidence   angle of reflection

shape, the object is distorted in a more complex way[2]. For example, parallel horizontal segments appear curved when reflected in a

cylindrical mirror, while parallel vertical segments appear to bow outwards, curves will appear straighter, and non-parallel segments will look parallel. The anamorphic image is retransformed back to its original shape only by using a cylindrical mirror.[3]

Positioning a cylindrical mirror just in the right plane will transform these straight segments into bowed ones, and vice versa.

Today computer technology can minimize some of the frustration and time expended on anamorphic art pieces. Check out the free online software *Anamorph Me!*[4] , the online-courses and classroom uses, and especially enjoy the online & off-line program QGoo (image morphing applet producing topologically equivalent distortions).

Whether it's creating an illusion or capturing reality, mathematics is behind the scenes. Mathematics may not have created the art or the illusion, but is invaluable in explaining why and how it takes place, and how to create your own anamorphic art.

---

[1] Today one comes across anamorphic art in many places. For example, check out Julian Beever's street art and the website www.anamorphosis.it.

[2] When light is reflected from a cylindrical mirror its angle of incidence and reflection are equal but measured from the tangent line to the curve of the mirror

[3] This type of anamorphic art was very popular as a parlor game and pastime during the 1800s. Today it can still be enjoyed in such books as *The Magic Mirror* by Dover Publications or on many online sites listed under *anamorphic art.*

[4] Go online to: myweb. tiscali.co.uk/artofanamorphosis /software.htm.

the sound of money

It sounds like money. Accentus' *Sonify!* software, a spinoff of technology developed at Dartmouth College, helps financial traders interpret quickly and accurately the constantly and rapidly evolving information that appears on

a recent 10 point move up. Traders easily learn the program's musical meanings by association with their normal work, and make connections between what the market is doing and the sounds they hear. Thus both the visual and auditory cortices of the brain are now on

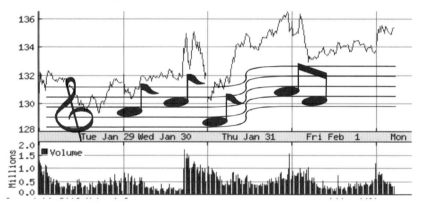

computer monitors by recognizing certain musical sounds. The software uses the auditory portion of the brain where the listener learns to interpret musical sounds which have been converted from the data streams. Depending on the user's musical acuity a trader may work simultaneously with several market indicators. Sonify! literally converts market changes to music. When these sounds are played, an adept trader can immediately recognize what economic changes are taking place. For example, the staccato notes of G, B, and C from a bassoon may indicate the Dow Jones is up 50 points on the day or

guard for financial traders, and enable them to literally make sound investments.

Some features and how they work:
Starting with a reference note, the user listens for changes in pitch (up/down indicating the market or a particular stock's respective movement). Certain sound clips or notes are alerts that a particular stock is approaching its target value or the market has undergone a major change. In addition to a note's pitch, it's loudness can also be used to indicate changes or movements.

witricity may cut the cord

W*itricity* is a term that refers to transmitting electricity wirelessly.

Will witricity do away with the clutter of cords and wires?

Although the concept is not new, scientists had dismissed its viability because they had no way to send the electromagnetic energy directly to a device without having it radiate in all directions.

Marin Soijacic, a physics professor at MIT, and his team(Aristeidis Karalis and John Joannopoulos) solved that problem by figuring how to have the device and the energy source tuned to resonate at the same wavelength. To demonstrate how this would work, the MIT team of researchers lit a 60-watt light bulb using power generated from an apparatus seven feet away not wired to the light bulb. The experiment also worked through wood barriers. This is just the first step in the direction of an office or home

without the clutter of cords and plug in charges, and, in addition, also reducing our dependency on batteries.

The team intends to work on scaling down the size of the transmitting device, increasing the range of transmission, and tweaking the method for efficiency. Presently only about 40% to 45% of the energy from the transmitting apparatus makes it to the light bulb; the rest remains near the emitting unit instead of being radiated and most loss is reabsorbed by the emitter. The researchers emphasize that the spill over produced by the magnetic coupling of the two devices is safe to human and electronic equipment. *"While rooted in well-known laws of physics, non-radiative energy transfer is a novel application no one seems to have pursued before.... figuring the details was not easy, Soljacic said, something he and his colleagues did through theoretical calculations and computer simulations."*[1]

In the not to distant future look for witricity to power your computer and laptops, cell phones and other house and office devices.

To further explore the science and mathematics used and developed by the MIT team to create witricity go to the following link: http://web.mit.edu/newsoffice/2007/wireless-0607.html.

---

[1]*Wireless Energy Transfer Can Potentially Recharge Laptops,Cell Phones Without Cords* by Davide Castelvecchi, published over the internet on November 14, 2006.

How do computers play chess? Methodically. Over the years scientists adapted machines and invented special techniques to use in computer chess programs. Among these are: *game trees, minimax algorithms, a bounded look ahead approach, static evaluation methods, alpha-beta pruning, and heuristics*. When the first move of the game is played the computer is instructed to generate all possible subsequent moves and counter moves leading to the possible outcomes of that game. Each *branch* of the *game tree* shows all the game boards with the possible moves and counter moves for a particular point in the game. At this point the computer must select the best move it can make. Here is where the algorithm *minimax* enters the picture. Theoretically the computer would start at the bottom final branch of the tree and assign each move a numerical value indicating whether the move is favorable for it or its opponent. Then, using this score, it would work its way

back up the branches to choose the best move for the stage of the game it is in. In chess there are far too many possible game boards for even

Fool's Mate, shown here, is one of the many checkmates that have earned names. Black was checkmated in just 5 moves from the game's start. Among other famous checkmates are: *Boden's Mate, Arabian Mate, Legall's Mate,* and *Greco's Mate.*

a computer to analyze. Here the computer program must use *a bounded look ahead approach*. This means that the computer instead of looking to the very last branch, only looks ahead maybe three or four branches. Then, the computer is instructed to use a process called *static evaluation* and assign a value for each game board up to this stage. Now the *minimax algorithm* kicks in and the computer works its way up the game tree to select the best choice for its next move. To speed up this process, programmers use *alpha-beta pruning* to instruct the computer to bypass (i.e. prune) unfavorable choices. Built into all this are *heuristics*— a self-learning approach to solve a problem by using feedback from actual performance to influence subsequent decisions. These heuristics may be in the form of inputted data bases of board configurations for good opening moves and endgame strategies. In addition, the computer can accumulate

clumsy and unsuccessful opponents. But as the computer's capacity and speed improved, programmers turned to harnessing the *brute force* capabilities of computers rather than relying on trying to have computers "think" their way through a game. The 1970s ushered in new more powerful machines and programs which harnessed this power.

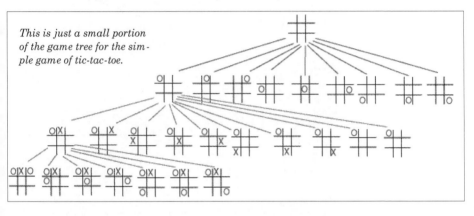

This is just a small portion of the game tree for the simple game of tic-tac-toe.

its own data bank from previously played games. Since a programmer selects the different moves, strategies and styles of play from the vast history of chess games and works of various grand masters, the programs end up reflecting the computer programmer's tastes, i.e. the programmer has left a fingerprint on the program.

Scientists initially tried to solve the "chess problem" by primarily focusing their efforts on artificial intelligence thereby trying to make the machine "think" like a human. These early computer programs failed to be a match for the human mind, and the computers were

By 1988 the first grand master was defeated by *Deep Thought,* the computer system produced at Carnegie Mellon University under the supervision of Feng-hsiung Hsu. Joining forces with IBM and introducing parallel computing techniques Hsu continued to work at the university, and *Deep Blue* was created. *Deep Blue* became the first computer to win a chess game against the reigning world champion Garry Kasparov.[1] Deep Blue returned in 1997, after undergoing a major makeover, and won a 6 game match against Kasparov.[2] Since then computer software and hardware have been continually improving to the point where computers are now expected to win matches against humans. Witness the Hydra vs Michael Adams match in June 2004. Hydra[2] showed its prowess by decisively defeating the 7th ranked chess player Adams. In a 7 game tournament, Hydra won 6 games and tied one. The computer made only one poor move, which came as a surprise, especially to its creator, Chilly Donniger.[3]

Deep Blue hardware.
Courtesy of IBM Corporate Archives

Deep Blue at play.
Courtesy of IBM Corporate Archives

Computers like humans have their Achilles heels. Kasparov discovered Deep Blue's weakness in the 1996 match, and although he lost the first game, Kasparov came back to

win the match. Would he have done the same in his 3rd match, if Deep Blue had accepted his challenge?

Today the world of chess competitions has a new face. No longer consisting of games of human vs human or human vs computer,

IBM's announcement of the Kasparov match in May 1997.
Courtesy of IBM Corporate Archives

computer vs computer matches now draw large audiences of chess aficionados. Just as world champions Garry Kasparov, Bobby Fisher, Anatoly Karpov, Mikhail Tal, and Boris Spassky are famous in the world of chess, today equally famous are digital champions such as Zappa, Hydra, Deep Blue, Shredder, and Junior.

---

[1]Although Kasparov lost the first game, he won the 6 game match—3 Kasparov, 2 tie, 1 Deep Blue.

[2]Deep Blue was now twice as fast as it was in 1996, capable of analyzing 200 million positions per second. It won 3.5 games of the match verses Kasporov's 2.5. Deep Blue consisted of a 30 node parallel computing system.

[3]Hydra, composed of a Linux cluster is based in Abu Dhabi, and plays in tournaments via the Internet. It is considered to be one of the strongest computer chess players in the world. Check its website http://www.hydrachess.com.

what powers mathematics?

What powers mathematics? Problems and imagination. Mathematicians are captivated by problems and puzzles. It may seem

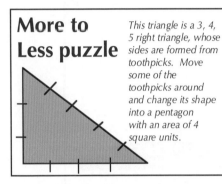

More to Less puzzle

This triangle is a 3, 4, 5 right triangle, whose sides are formed from toothpicks. Move some of the toothpicks around and change its shape into a pentagon with an area of 4 square units.

like an obsession to a layperson; but to the mathematician, it is a challenge, no different than a sculptor challenged by a piece of marble, or a chef by a new recipe. In 1900, mathematician David Hilbert delivered a lecture to the Second International Congress of Mathematicians in Paris. He discussed the importance of mathematical problems, saying *"As long as a branch of science offers an abundance of problems, so long is it alive; a lack of problems foreshadows extinction or cessation on independent development. Just as every human undertaking pursues certain objects, so also mathematical research requires its problems."[1]* At this time, Hilbert presented a list of 23 mathematical

problems he hoped would be challenging during the 20th century.[2] Today mathematics is definitely not short of problems. Search the Internet for mathematics problems—there are literally thousands of problems posted.[3]

Without imagination, new innovative solutions would not be possible. Without imagination, mathematics would not exist. As mathematicians seek a solution to a problem, a journey begins whose destination is unknown until the mathematical mind follows the path of logic to its end. History demonstrates time and again that the mathematical journey is not merely reaching the solution but the discovery of new landscape along the way. For example—

• Mathematicians in their efforts to to prove Euclid's Parallel postulate discovered new non-Euclidean geometries.

• The three impossible problems of antiquity led ancient mathematicians beyond the compass and the ruler to explore conics in depth and to develop ingenious methods and tools such as the cissoid of Diocles, Plato's cube doubler, and the conchoid of Nicomedes.

Whose bill is this?

Tom, Dick, and Harry have been friends for over 20 years. One is a habitual liar. One sometimes fabricates a lie. One is scrupulous about telling the truth. One of them dropped a $20 bill by their feet. If the $20 bill belongs to the liar, then who is this, if:

- *Tom says: The $20 is mine.*
- *Dick says: Tom is telling the truth.*
- *Harry says: It belongs to Dick.*

• Euler's solution to the Königsberg bridge problem (page 105) not only showed that crossing all seven bridges was not possible but also introduced the use of networks to problem solving and helped launch the field of topology.

• Fermat's Last Theorem (see page 105) took centuries of mathematical work. Along this journey, mathematicians Euler, Gauss, Germain, and Kummer developed sophisticated methods in number theory. André Weil, Yutaka Taniyama, Gerhard Frey, Gerd Faltings, Goro Shimura, Kenneth Ribet, Barry Mazur, and Andrew Wiles explored such diverse areas as elliptical modular curves, modular forms, and the Taniyama-Shimura conjecture until FLT was finally proven by Andrew Wiles in collaboration with Richard L. Taylor.

Try out your logic on the three problems in this section.

Stacking squares

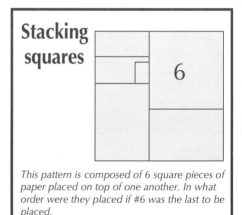

This pattern is composed of 6 square pieces of paper placed on top of one another. In what order were they placed if #6 was the last to be placed.

---

[1] From David Hilbert's address to the International Congress of Mathematicians 1900, it was translated into English by Dr. Mary Winton Newson for Bulletin of the American Mathematical Society 8 (1902), 437-479. A reprint appears in Mathematical Developments Arising from Hilbert Problems, edited by Felix Brouder, American Mathematical Society, 1976.

[2] Several of the 23 problems still remain unsolved today.

[3] Some websites to explore are http://www.mathworld.wolfram.com, 20,000 problems under the sea at http://www.problems.math.umr.edu, and http://www.claymath.org.

some things don't change
—euler's polyhedra theorem

$\mathbf{W}$hether it's a cube, an octahedron, or, in fact, any polyhedra[1], if you add the number of its faces to the number of its vertices and then subtract the number of its edges from this sum, the result is always 2.

Swiss mathematician Leonhard Euler (1707-83) discovered this polyhedron characteristic in the mid-1700s, and it has been proven in many ways using different areas of mathematics includ-ing graph theory, in-duction, manifolds, to-pology. Each new proof may unveil new mathematical ideas or new uses for old ideas. There are even proofs showing it holds true for higher dimensional forms.

A dodecagon—a 12 faced polyhedron with pentagon faces.

If you have never test-ed out Euler's polyhe-dra theorem, try it out on such shapes as box, pyra-mid or any solid poly-hedron you have around.

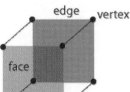

face   edge   vertex

Here's an informal explanation of why Euler's polyhedra theorem holds true. We will take a cube and change its form by either adding a vertex or an edge, and notice how/if the sum of **V+F-E=2** changes.

By *adding a new vertex point,* we lose the shaded face because it now becomes part of the solid's interior. The new vertex also adds 4 new faces, but keep in mind the shaded face is lost. In addition, 4 new edges also appear. The net gain of *new vertices, new faces* and *new edges* is 1+3-4=0,

meaning no change in the original value of **2**. *Extend this reasoning to any n-sided face solid, and add a vertex as we did with the cube's face.* **n** *new faces will be formed, but one will be lost, so the net gain of faces is* **(n-1)**. **n** *new edges are formed. Thus, there is no net change because* **1+(n−1) − n=0** .

Now consider *adding a new edge* to the cube. The solid looks like this. Every new edge has 2 new

vertices. There are 4 new faces but the interior shaded face is lost, so the net gain of faces is 3. There are 5 new edges. Summing the new added parts gives 2+5 − 5=0, and again there is no change in the original value of **2**. *Now add an edge to an n-sided face solid, and determine that there will be no net change for V+F−E because* **2+(n-1) − (n+1)=0.**

---

[1]Closed solids whose faces are polygons.

mathematics unravels quasicrystals

Quasicrystals, as their name implies, are not your everyday crystals. In fact, they did not even exist until 1982 when chemist Daniel Shechtmann made a new super strong alloy formed from aluminum and manganese. Up until then, crystallographers relied on the mathematics of tessellations and a mathematical theorem (Barlow's theorem) to describe the structure of crystals. They assumed crystals were composed of *periodic* arrangements of a tessellated polyhedra whose patterns possessed *rotational symmetry of three, four or six-fold symmetry.* This meant that the polyhedra building blocks fit together so that when they were rotated either 1/3, 1/4, or a 1/6 of a circle the pattern looked exactly like it did before turning it.

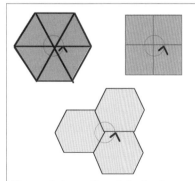

The tessellation or tiling of triangles above possess six-fold symmtry because its tessellated pattern looks exactly the same six times in one complete turn. In other words a sixth of a turn does not change its pattern. Similarly the squares' tiling has four-fold symmetry, while regular hexgons' have three-fold symmetry.

Yet Shechtmann's alloy was not periodic nor possessed one of the conventional rotational symmetries of

A cluster of a normal crystal.

crystals. Shechtmann was pressed to convince fellow scientists that this was indeed the case, and that this alloy had five-fold symmetry. It was not until 1984 that physicist Paul Steinherdt and then graduate student Don Levine, using computer simulation created a virtual crystal, which Steinherdt dubbed a *quasicrystal.* This simulation exactly described Shechtmann's alloy which was composed of non-periodic tiles and possessed five-fold symmetry. Solid matter can be described as amorphous (without shape) or crystalline. In an amorphous solid, the atoms are randomly arranged with no pattern. On the other hand, a crystal's atoms are arranged with periodic order to them. With Shecthmann's discovery, crystallographers realized that certain material could also have five-fold symmetry. These discoveries forced crystallographers to redefine crystals to include quasicrystals. The International Union of Crystallography redefined the term *crystal* as any solid having an essentially discrete diffraction diagram produced when beams of x-rays or electrons are scattered by atoms of a solid.

Since Shechtmann formed his alloy, over a hundred new quasicrystals have been discovered. Some even possess 7-fold and 9-fold symmetry, yet their structures remain a mystery. Scientists and mathematicians are trying to unravel quasicrystals by applying concepts from such mathematical areas as Fouier space, lattice theory, Penrose tilings, wave vectors, statistical distributions, computer simulations.

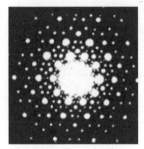

The patterns illustrating the atomic structure of a solid are the arrangement of white spots on a dark background of the x-ray film. If the spots, unlike this diagram, are fuzzy, i.e. not discrete, the solid is glass rather than crystal.

Mathematician David Damanik from Rice University, having published a key proof about quasicrystals in 2007, says *"Mathematically speaking, quasicrystals fall into a middle ground between order and disorder. Over the past decade, it's become increasingly clear that the mathematical tools that people have used for decades to predict the electronic properties of materials will not work in this middle ground."*[1] Mathematicians have shown that quasicrystals are too complex to use what's known as Schrödinger's operator[2] to determine if the quasicrystal will be a conductor or insulator. Thus far mathematicians have proved that past methods for determining if material will be a conductor or an insulator do not work for quasicrystals. Daminik further explains that *"Computer simulations have shown that electrons move through quasicystals—albeit very slowly—in a way that's markedly different from the way they move through a conductor. But computers never show you the whole picture. They only approximate a solution for a finite time. In our paper, we proved that electrons always behave this way in the quasicrystal model we studied, not just now or tomorrow but for all time."* Beware quasicrystals! Mathematicians are out to discover and prove just what makes you tick.

---

1 *Quasicrystals: Somewhere between order and disorder* from *Science Daily*, May 29, 2007.

2 Up until quasicrystals, Schrödinger's operator from the Schrödinger's equation was used to find out how electrons behave in any material. It was the work of Austrian physicist Erwin Schrödinger that in 1926 described the space-time-dependence of quantum mechanical systems.

nicomedes' conchoid goes green

Now, over 2100 years later, the ancient conchoid of Nicomedes has joined forces with the virtual sunflower to tap into new ways to harvest solar energy. *Energy Innovations* is concentrating its attention on its innovative use of the conchoid for its solar concentrator system called the *Sunflower*. Just as sunflowers constantly turn toward the Sun and follow its path throughout the day, the *Sunflower*, a system of 25 mirrors controlled by only two motors, is able to adjust each of these mirrors to constantly recreate the conchoid and thereby concentrate and collect the rays' energy through its PV panel everyday throughout the year. Energy Innovations is currently designing these units for rooftops, where they can occupy unused space unobtrusively. More importantly the units are self-contained and thereby eliminate power companies and power brokers.

A single Sunflower unit. Photographs courtesy of Energy Innovations.

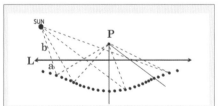

A rendering of how a group of Sunflowers would be linked into an integrated rooftop system.

If the conchoid is inverted, so pole P is on top as shown in the diagram above and at all the points of the conchoid there are mirrors positioned so they emulate the shape of the conchoid, then as the Sun's rays hit these mirrors the rays are reflected and bounce to point P. If at point P there is a small panel of solar cells known as PV cells(photovoltaic cells), the concentrated light beam can be converted to electricity. Until the conchoid, the curve of choice was the parabolic curve. In fact, huge parabolic troughs and other shaped dishes have been developed over the years.

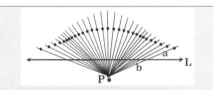

To form a conchoid begin with a line L and point P. Then draw rays through P which intersect L. Mark off a fixed distance, **a**, on each of these rays. The locus of these points form the conchoid. The curvature of the conchoid depends on the relationship between a and b, i.e. a=b, a<b, or a>b. The polar equation of the conchoid is r=a+bsecq .

constants, constants everywhere

Pick a number, any number, between 0 and 10. There are millions, trillions...an infinite number of mathematical constants from which to choose. But, most people usually choose 1,2,3,4,5,6,7,8 or 9. Why? Throughout our lives we have been conditioned to think in whole numbers. Few people will choose e, phi or pi—all famous and all between 0 and 10.

Over the millennia the diversity of numbers entered the world of mathematics mainly through solutions to problems. Problems would pop up whose answers were not found in the existing set of known constants. For example, 7 taken away from 5 may have been a type of problem that brought about the necessity for negative numbers. And even though negative numbers were introduced to Europe by merchants who had contact with Arabs who used negative numbers, it took decades for mathematicians to accept them as viable numbers. Consider the simple looking equation, $x^2 = -1$. Its solutions could only be found by introducing a new set of numbers—the imaginary numbers. Geometric and algebraic problems gave birth to such numbers as $\sqrt{2}$, -3, 3.14159..., the irrational numbers, the transcendental numbers, the complex.

Patterns also had a lot to do with the discovery of various constants. In the mid-1970s physicist Mitchell Feigenbaum, working with iterative equations, noticed how a particular constant (approximately 4.449211660...) appeared again and again in his calculations. In the process he uncovered a *universal constant,* a number common among a certain class. In this case it was the class of chaotic systems, similar to how pi is a universal constant for the class of geometric objects called circles. This constant is linked to re-iterative systems, and is now known as *Feigenbaum's constant* and is designated by the Greek letter delta, δ. These and countless others are special constants, and are usually associated with a class of objects with which they always occur. Look at the class of all squares. Regardless of its size every square's diagonal is always equal to the $\sqrt{2}$ times the length of its side. The same, as mentioned above, is true for the class of circles because any circle's circumference is always equal to pi times its diameter.

Over the years as new special constants were discovered, in addition to their number name, they usually were named after their discoverers and given a symbol name, especially when their number names are never ending non-repeating decimals. Consider the number

1.618033988749894... It's the famous *golden ratio* in decimal form. It could also be written as $(1+\sqrt{5})/2$, but instead it's been given the symbol, ϕ, phi (named after the famous

was reflected in the sequence of numbers 1,1,2,3,5,8,13,...in which each term is the sum of the two previous terms. *The Fibonacci numbers,* as this sequence is now called,

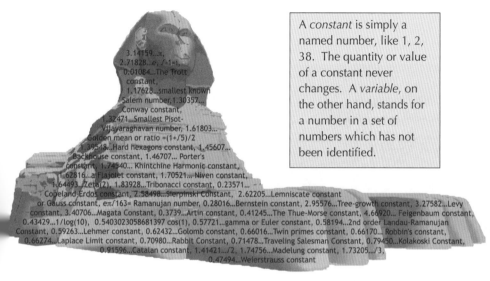

3.14159...π,
2.71828...e, $\sqrt{-1}=i$,
0.01084...The Trott constant,
1.17628...smallest known Salem number, 1.30357...
Conway constant,
1.32471...Smallest Pisot-Vijayaraghavan number, 1.61803...
golden mean or ratio $=(1+\sqrt{5})/2$
1.39548...Hard hexagons constant, 1.45607...
Backhouse constant, 1.46707... Porter's constant, 1.74540... Khintchine Harmonic constant,
1.62816...a Flajolet constant, 1.70521... Niven constant,
1.64493...Zeta(2), 1.83928...Tribonacci constant, 0.23571... -
Copeland-Erdos constant, 2.58498...Sierpinski Constant, 2.62205...Lemniscate constant
or Gauss constant, $e\pi/163=$ Ramanujan number, 0.28016...Bernstein constant, 2.95576...Tree-growth constant, 3.27582...Levy constant, 3.40706...Magata Constant, 0.3739...Artin constant, 0.41245...The Thue-Morse constant, 4.66920... Feigenbaum constant,
0.43429...1/log(10), 0.5403023058681397 cos(1), 0.57721...gamma or Euler constant, 0.58194...2nd order Landau-Ramanujan
Constant, 0.59263...Lehmer constant, 0.62432...Golomb constant, 0.66016...Twin primes constant, 0.66170... Robbin's constant,
0.66274...Laplace Limit constant, 0.70980...Rabbit Constant, 0.71478...Traveling Salesman Constant, 0.79450...Kolakoski Constant,
0.91596...Catalan constant, 1.41421.../2, 1.74756...Madelung constant, 1.73205.../3,
0.47494...Weierstrass constant

A *constant* is simply a named number, like 1, 2, 38. The quantity or value of a constant never changes. A *variable*, on the other hand, stands for a number in a set of numbers which has not been identified.

Greek sculptor Phidias who supposedly used phi in the proportions of his work). The number 2.718281828459045... was uncovered when logarithms were discovered in the 1600s, and did not get its symbol name until Euler used **e** (probably derived from the word exponential) for its symbol in the 1700s. In addition, Euler was responsible for giving 3.415926... its symbol name π more than 5000 years after it was discovered. When Leonardo da Pisa (aka Fibonacci) wrote a problem about a group of reproducing imaginary rabbits, a sequence of special constants surfaced. The answer to his problem

are usually signified by f_s, such as f_1, f_2, f_3, f_4,..... Over the centuries Fibonacci numbers have been found to occur in many facets of nature, such as the swirls of pine cones, the growth patterns of certain twigs, trees, and the seedheads of certain flowers to name a few. This sequence of numbers is also connected to many mathematical concepts[1]. For example, the ratio of consecutive Fibonacci numbers gets closer and closer to the value of the golden ratio.

In 1999 computer scientist Divakar Viswanath discovered a new special constant that also relates to the Fibonacci numbers. Out of curiosity

Viswanath decided to see what would happen if he introduced an element of randomness when generating a Fibonacci-like sequence. Starting with the first two Fibonacci numbers, 1 and 1, instead of adding these two terms to get the next

1. 1 is its own symbol name, but no less important or interesting than any of the others. It's also very special—it's the only number that is a factor of every other number; it's one of the two numbers used in the binary number system; it's the only

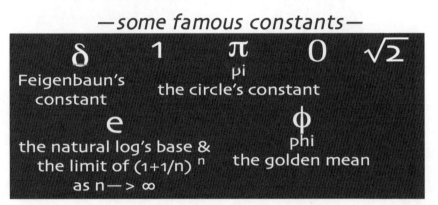

—some famous constants—

δ Feigenbaun's constant

1

π pi the circle's constant

0

√2

e the natural log's base & the limit of $(1+1/n)^n$ as $n \longrightarrow \infty$

φ phi the golden mean

term, he flipped a coin. If it came up heads he added the two previous terms. If it came up tails he subtracted the two and dropped the negative sign if it appeared (i.e. he took the result's absolute value as the new term). He found that regardless of what Fibonacci-like sequence this random process generated, the ratio of its two consecutive terms continually got closer and closer to the constant 1.13198824... . This special constant is so new it does not as yet have a symbol name, but is referred to as the *Viswanath constant*.

Bear in mind, in the group of special constants we even have everyday constants such as the number

number whose reciprocal is itself; it's the first counting number; it's the only number that does not affect any number's value when it multiplies or divides that number; and it was the first number ever written!

The universe abounds with numerical constants. Have they always existed? Or is their existence dependent on mathematics? Although it seems each new constant appeared to help communicate an idea, answer a question or solve a problem, one wonders if constants are embedded in the fabric of the universe, or do humans introduce constants in order to describe, understand and explore nature?

---

1 The Fibonacci sequence is connected to the Pascal triangle, the golden ratio, the golden rectangle, mathematical identities and mathematical tricks..

a mathematical oasis

The Mathematical Sciences Research Institute, perched in the Berkeley Hills on Gauss Way, affords magnificent views of the San Francisco Bay Area and is just minutes away from University of California Berkeley Campus. Founded in 1982 as a nonprofit institute, MSRI is a haven for mathematicians. It is designed especially to be conducive to mathematical research. As its mission statement points out *"Dedicated to the advancement and communication of fundamental knowledge in mathematics and the mathematical sciences, to the development of human capital for the growth and use of such knowledge, and to the cultivation in the larger society of awareness and appreciation of the beauty, power and importance of mathematical ideas and ways of understanding the world."* It provides participants both isolation and interaction. It facilitates as a sounding board for mathematicians to share new ideas and hash over olds ones. Even the institute's grounds host art that speaks mathematics. In its patio garden we find a sculpture by Helaman Ferguson called *The Eightfold Way,* which represents the Klein Quartic. A wall is gracefully tessellated with colored tiles forming an aperiodic tiling of rhombi, squares, rectangles, hexagons, stars. Many of the tiles themselves have mathematical messages. Even a glass window has Karl Gauss' construction of the regular 17-gon.

Chalkboards are strategically placed so that casual conversation and discussions can flow easily into equations and formulas onto them.

Gauss' 17-gon construction at MSRI.

MSRI also embraces and includes public participation in lectures and programs, such as M*A*T*H which featured Alan Alda in conversation with Bob Osserman in January 2008. The MSRI website is a treasure trove of calendar event information and even downloadable on-line lectures.

Looking to the future, MSRI hosts an undergraduate program, MSRI-UP, which includes six summer weeks at the institute where undergrad students can conduct research and do mathematical research presentations. The program also includes a network of mentors and follow-up support.

In addition to using the UC Berkeley libraries, the MSRI has its own mathematics library of over 20,000 books and journals, and also publishes papers and books annually.

To learn more about MSRI and its events go to: www.msri.org/.

4 deceptively simple blockbuster problems

Here are four deceptively simple problems whose proofs have thus far eluded mathematicians.

GOLDBACH'S CONJECTURE
—1742—

Can every even number greater than 2 be written as the sum of two primes? On the surface, the layperson wonders "what's the big problem?"...but when a proof is attempted, it may take additional centuries to reach a solution.

In 1742, mathematician Christian Goldbach wrote in a letter to Leonhard Euler that he believed that *every integer greater than 5 is the sum of three primes*. Euler replied suggesting that Goldbach's statement was equivalent to the statement that *every integer other than 2 is the sum of two primes*. The conjecture in this form was first published in *Meditationes algebraicae* in 1770, and came to be called *Goldbach's Conjecture*. The *Goldbach Conjecture* has a prominent role in the novel, *Uncle Petros and Goldbach's Conjecture* by Apostolos Doxiadis. The book was published in 2000, and its publisher offered a million dollar prize for the conjecture's proof, provided that it was submitted prior to April 2002. Various proofs of different levels of the problem were presented, but the Goldbach Conjecture remains elusive.

THE ODD PERFECT PRIME PROBLEM
—4th century BCE—

Does an odd perfect prime number exist? A perfect number is one which is equal to the sum of its proper divisors **I'm perfect** **Well, so am I!**

6 **28**

(a divisor other than the number itself). 6 is perfect because 6=1+2+3.

THE TWIN PRIME CONJECTURE
— 18th century —

Is there an infinite number of twin primes? Two primes are twins if their difference is two.

You don't look like my twin.

5 **7**

(For recent work on distribution of primes, see website http://www.olimu.com/Notes/Goldst onYildirim.htm.)

THE PERFECT CUBOID PROBLEM
— 1719 —

A cuboid is a hexahedron, which is a brick or box shape. It's a *perfect cuboid* if all its edges and all its diagonals are integers. Does a perfect cuboid exist? None have been found, and no one has been able to prove none exist.

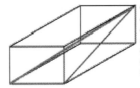

One of this cuboid's interior diagonals is shown in blue and one of its facial diagonals is shown in red.

the turing test

In 1950, the brilliant mathematician and computer theorist Alan Turing posed the question "Can Machines think?" in his work *Computing and Intelligence.* In this article he discusses many areas including definition of the words "machine" and "think", scenarios of possible imitation games to test his question, digital computers, theology, mathematics and consciousness.

One side of the bronze metal shows Alan Turing and the other has Hugh Loebner, who had underwritten the award.

He proposed that, *"... in about fifty years' time it will be possible, to programme computers... to make them play the imitation game so well that an average interrogator will not have more than 70 per cent chance of making the right identification after five minutes of questioning."*

Today the Loebner Prize for artificial intelligence is an annual contest aimed at creating and implementing the Turing Test. Each year a prize of $2000 and a bronze medal is awarded to the most human-like computer in comparision to the other entries. The 2008 contest will be held at Reading University, UK. For additional information on the contest and past winners go to http://www.loebner.net/Prizef/loebner-prize.html.

Turing's imitation game

Turing describes the imitation game as a game "played with three people: a man (A), a woman (B), and an interrogator (C) who may be of either sex. The interrogator stays in a room apart front the other two. The object of the game for the interrogator is to determine which of the other two is the man and which is the woman." Turing discusses this and then asks *"What will happen when a machine takes the part of A in this game?"* To read his entire article go to: http://www.loebner.net/Prizef/TuringArticle.html

the four forces of the universe

Imagine the mathematical work that went into developing such famous equations as Albert Einstein's $E=mc^2$ or Michael Faraday's law of electromagnetic induction, $\nabla x E = -\partial B / \partial t$. Physicists have to be well versed in mathematics to deal with equations and formulas that

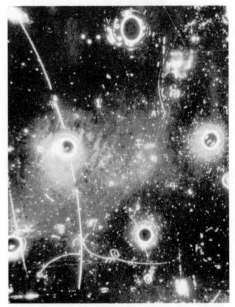

At CERN in July 1973, a test produced particle paths revealing the ground breaking discovery which confirmed the electroweak theory showing that the weak force and the electromagnetic force were different facets of the same interaction. Photo © CERN, all rights reserved by CERN. Photo courtesy of CERN.

deal with the four forces that affect our world. Just understanding these forces is a difficult task, let alone working with the mathematics that defines them.

Forces, also called interactions, can be thought of as the stuff that binds matter. Gravity may not seem like a weak force to us especially when we try to get up from a supine position, but for matter the size of particles, gravity is very weak. The larger the size of the matter (e.g. planets) the stronger the force (or pull) of gravity. *Gravity* is a force that attracts two objects.

Electromagnetism, on the other hand, can both attract and repulse two objects. Scientists noticed that the further away an object is from gravity or from electromagnetism the weaker is this force's affect on the object. Inside the atom, things are different. The *weak (nuclear) force* works on decaying neutrons such as in radioactivity, and *weak force* also changes protons into neutrons. The *strong (nuclear) force* holds quarks inside protons, neutrons and other particles. Inside the nucleus it keeps protons from flying off if they come under the repulsive action of electromagnetism because protons all have positive charges. The *strong force* is 100 times stronger than electromagnetism inside the nucleus. And unlike electromagnetism and gravity, the *strong force* becomes even stronger as distance between objects increases, which explains why tremendous amounts of energy are needed to try to separate particles.

In the future look to the field of nanotechnology for the design of cables of amazing strength made of nanowires formed from carbon nanotubes. Yes, this idea is real and these nanotubes are not only envisioned for cables, but for the *space bridge,* also called *space elevator* or *beanstalk.* The space bridge would connect a planet (e.g. Earth) with a point in outer space. The Earth and a point 35,786 km above can be theoretically bridged by an elevator whose center of mass is at the Earth's geostationary point with a 24 hour orbit which allows it to remain stationary over the Earth's equator as the Earth rotates. The space bridge would be a conduit used to transport materials and people thereby minimizing the use of rocket propulsion. The space bridge will be another example of where imagination, engineering, architecture and mathematics join forces to create amazing architectural wonders.

This rendition of a space elevator, conceived by artist Pat Rawling, looks down along the geostationary transfer station toward Earth. Courtesy of NASA image gallery.

Technically, anything between the size of 0.1 and 100 nanometers can be considered a nano-object.

A nanometer is one-billionth of a meter, 1/1,000,000,000 meter.

2 is odd

2 is not one of your everyday integers. It's been around and recognized for thousands of years.

2 is definitely not an an *odd number,* but it certainly has some distinctive characteristics and appears in some very diverse places.

• 2 is the smallest even natural number. It's also the very first prime number and the only even prime number. The Pythagoreans were not even sure it was a number because it had a beginning and end, but no middle.

• 2 is the only number which when added to itself gives the same result as when multiplied by itself. $2+2 = 2 \times 2$

• A number is divisible by 2 if its ones (the units) digit is divisible by 2.

• Other than 1, 2 is the only other natural number whose factorial equals itself, $2! = 1 \cdot 2 = 2$

• A number is divisible by 2^n (a power of 2) if the last n digits are divisible by 2. For example, 7458120 is divisible by 2^3 because 2^3, namely 8, divides into the last three digits of this number, i.e. 120. 2^5 or 32 divides evenly into 390146912 because 2^5 goes into the last 5 digits, (namely 46912) of 390146912

• 2 is not only used to describe any even number, but also any odd number. For the even numbers the formula is 2n, where n is an integer. For the odd numbers the formula is 2n+1. For example, to write 35 using the 2's odd formula, divide 35 by 2, we get 17 with remainder 1, so 35 can be written as $2 \cdot (17)+1$.

• 2's radical , $\sqrt{2}$, was the first recorded irrational number. It always appears in a square, either on its diagonal or its sides. An example is:

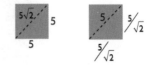

• A set consisting of n distinct elements always has 2^n subsets. For example, {a, b, c, d} has 2^4 or 16 subsets, namely {a}, {b}, {c}, {d}, {a,b}, {a,c}, {a,d}, {b,c}, {b,d}, {c,d}, {a,b,c}, {a,b,d}, {a,c,d}, {b,c,d}, {a,b,c,d}, { }. A total of 2^4 or 16.

• 2 also appears in the rows of the Pascal triangle. The sum of the digits in any row of the Pascal triangle is a power of 2, namely the nth row's sum is 2^n. For example *row 5* numbers can be totaled by

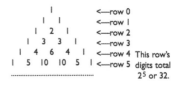

simply taking 2 to the 5th power, namely $2^5 = 32$.

• If *Goldbach's Conjecture* is true, 2 is the only even number that can't be written as the sum of two primes. For example, 6=3+3; 8=5+3; 28=23+5

• *Any* integer is the sum of a sequence of consecutive integers *if and only if* the integer is not a power of 2. For example, 85 is not a power of 2, therefore there is a group of consecutive integers that total 85, namely 15+16+17+18+19. But you won't find a sum of consecutive integers for 2, 4, 8, 16, 32, 128, 256, ...

• 2 is the base number for the smallest place value system for writing natural numbers. Called the *binary system* it uses the digits 0 and 1 to write any integer. For example, 21 is written as 10101, where $1(16=2^4)+0(8=2^3)+1(4=2^2)+0(2=2^1)+1(2^0=1)$. The binary system is the one used by electronic computers because 1 represents electricity "on" and 0 is for "off".

• 2 makes its presence known in some well known theorems.

—*Fermat's Last Theorem* has a 2 connection. The theorem states: *There are no positive whole numbers—x, y, z — that solve $x^n+y^n+z^n$, when n is a natural number greater than 2.*

—Mathematicians came up with a spin off of Fermat's Last Theorem, namely the equation was rewritten so that $n^x+n^y=n^z$. This equation's *only integral* solution is $2^1+2^1=2^2$, which is a spin off of 2's property that $2+2 = 2 \times 2$.

• The number 2 appears prominently in two of Leonhard Euler's discoveries.

—*Euler's polyhedra theorem* shows that the number of vertices V, the number of edges E, the number of faces F for any simple closed polyhedron are always related to the number 2 as follows: **V+F–E=2**. For example, a cube has 8 vertices, 6 faces and, 12 edges. Thus V+F–E =8+6 –12 = **2**.

—The *Königsberg bridge problem*. When Euler drew a network illustrating the Königsberg bridge problem, he proved that the only way it would be traceable (i.e. traveled so each path is used only once) was if the network had only 2 odd vertices. Euler reasoned for a vertex to be odd the path would either begin or end at that vertex. Consequently, since a traceable path has only one beginning and one ending, it could have only 2 odd vertices. The diagram above shows the network for the seven bridges and paths of the Königsberg bridge problem. Notice all vertices for the network are odd.

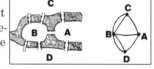

Yes, 2 is very special, unusual, and rather odd.

hacking into communication of bacteria

It's mind boggling to learn that different species of bacteria communicate. The story of this discovery begins with the *Vibrio fischeri,* a bacteria that inhabits the bodies of certain squid and jellyfish and at times glows or lights up their bodies. These microbes secrete certain chemical molecules, called *autoinducers.* When the concentration of

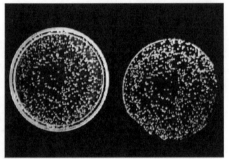

Vibrio fischeri colonies growing on agar. Left photo was taken with a light source, while the one on the right was taken using the colonies own bioluminescence as a light source. Courtesy: National Science Foundation and J. W. Hastings, Harvard University, through E. G. Ruby, University of Hawaii

certain autoinducers in the squid or jellyfish reach a certain level, these bacteria react by glowing. The autoinducers help them keep track of their numbers. When they sense their population has reached a certain number, they react in unison. This phenomenon is called *quorum sensing.* In the case of the Vibrio fischeri, they react by glowing. Bacteria's *quorum sensing* mechanism triggers varied responses in different species. For example, if *cholerae,* the bacteria which causes chol-

era, is present in a human's stomach its quorum sensing tells them when to release toxins. It also alerts them when to stop producing toxins and begin excreting an enzyme that will release them from the intestine to move out of the body and seek other people to infect.

Working with this glowing phenomenon of the Vibrio fiseri, biologist Bonnie Bassler discovered that two communications systems were actually used by the bacteria. One which the bacteria use to communicate among themselves and another they use to communicate with different species of bacteria. She discovered this phenomenon when she curtailed the autoinducer with which the bacteria communicated with one another. With this autoinducer absent, it made sense that they should have stopped glowing. Yet, these bacteria continued to glow because foreign bacteria in their vicinity were secreting another autoinducer, named AI-2, which the Vibrio bacteria understood to mean that a critical mass of bacteria was still in their vicinity. Working with these findings, Bassler and her team of researchers were able to isolate a gene called LuxS which was responsible for producing the AI-2 autoinducer. Furthermore, they found that that this gene is present in the hundreds of bacteria they studied. In other words, Bassler's team hit upon a language (A1-2) that is universally understood by different species of bacteria, even though each species has its own language. Bassler points out *"Bacteria can communicate between species, and they have evolved mechanisms to interfere with the communication. Probably this is but one of many cunning strategies they have for manipulating chemical communication."*[1]

A beneficial bacteria, *E. choli,* which lives in the human stomach, has the ability to inhibit quorum sensing mechanism of virulent bacteria such as the *cholerae* by consuming the A1-2 autoinducers and fooling the virulent bacteria into not knowing they have a quorum. Virulent bacteria are careful not to act before a critical mass is reached, otherwise the human immune system could attack and destroy their meager numbers. Quorum sensing is crucial for bacteria survival, but it also plays an important role in preventing bacteria from acting as a unified large "multi-celled organism". Bassler explains that *"The real take-home point is the interference. Consumption of the signal could be a mechanism that allows one kind of bacteria to block another kind of bacteria from counting how many neighbors they have and, in turn, properly controlling its behavior."*[2]

She further adds *"...that understanding quorum sensing in bacteria will allow us to learn the principles underpinning multicellularity in all organism."*[3] Bassler was awarded the MacArthur Foundation *genius grant* in 2005 in recognition of her amazing work. She and her research team continue to make use of a spectrum of tools including those used in genetics, biochemistry structural biology, chemistry, microarray studies, bioinformatics, and modeling to explore, explain and understand bacterial communication and cell signaling in higher organisms and to discover new ways to combat bacterial diseases.

Bioinformatics is a relatively new field of science that encompasses many fields of mathematics in the study of biology. Included are computer science, artificial intelligence, cellular automata, statistics and applied mathematics. With the advent of genome sequencing came hugh databases to store, access and to add biological information. Analyzing, identifying, and interpretating such vast amounts of information dealing with genes, nucleotide and amino acid sequences, proteins' domain and structure, all fall under the field referred to as *computational biology*. Locating a gene within a sequence, predicting protein functions and structures, and pinpointing related sequences often require the development of mathematical formulas, data mining techniques, and statistical anaylsis.

---

[1] From *Say What? Bacterial Conversation-Stoppers*. Howard Hughes Medical Institute *Research News*, September 29, 2005.

[2] Ibid.

[3] From *Small Talk with Bonnie Bassler* by Carmen Drahl, Research Advances, AWIS Magazine, Volume 35, November 1, 2006.

branes, gravity, and all that stuff

Physicists have never been at a loss for ideas about the Universe. Today, among their theories are such notions as multiple dimensions, parallel universes, minuscule vibrating strings defining matter, supersymmetry and dark matter and energy. Physicists Lisa Randall and Raman Sundrum have drawn from these ideas and established physics to create a theory that encompasses and works with what we know thus far. In the Universe they see our world existing on a *brane*. We reside on the brane in 3-dimensions (length, width, thickness), and the particles of matter and their interactions (aka forces) compose everything on the brane. Like other string theories, particles' basic units are infinitesimally small vibrating strings whose vibrations determine the composition of its matter and energy. As in the Standard Model of particle physics the four *forces*(the strong nuclear, weak nuclear, electromagnetism and gravity) interact on our brane. Our brane, along with perhaps other branes in the Universe, float in what is called *bulk*. Bulk is the term used to describe the full higher-dimensional infinite space spreading out in all directions of the Universe. Branes are membranes of lower dimensions than the surrounding bulk. Life on our

brane is the same as it's always been on our world. We and all matter experience the *four forces* on our brane. Gravity, the force that makes objects move toward one another, remains a very weak force, so weak that it cannot even pull a paper clip away from a small magnet. *Why is gravity so weak?* Science does not know for sure, but the *Randall-Sundrum theory* has a possible answer. Gravity, unlike the other three forces on our brane, does not actually reside on our brane. It is not a captive of our brane. Apparently, unlike our other three forces, gravity communicates or travels throughout bulk via

---

Standard Model refers to the present day accepted theory that has been experimentally developed for particle physics. It includes the four non-gravitational forces (weak nuclear, strong nuclear, and electromagnetism) and the particles and subatomic particles thus far discovered and how they interact according to established physical laws.

The Randall-Sundrum model considers a *graviton particle* to be a *closed string*, meaning it has no ends. *Open strings*, have two ends, must have their ends on branes— either the same brane or separate branes. In other words these open strings particles are confined to a brane or two branes, while a graviton is free to travel through the bulk.

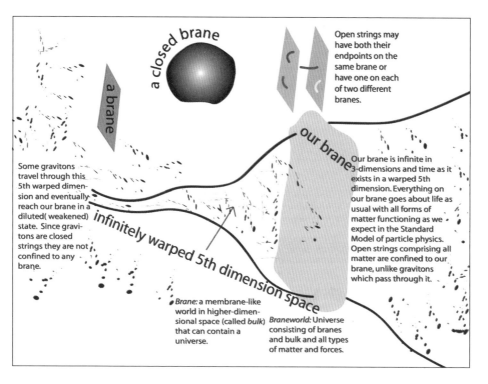

a closed brane

a brane

Open strings may have both their endpoints on the same brane or have one on each of two different branes.

our brane

Some gravitons travel through this 5th warped dimension and eventually reach our brane in a diluted(weakened) state. Since gravitons are closed strings they are not confined to any brane.

infinitely warped 5th dimension space

Our brane is infinite in 3-dimensions and time as it exists in a warped 5th dimension. Everything on our brane goes about life as usual with all forms of matter functioning as we expect in the Standard Model of particle physics. Open strings comprising all matter are confined to our brane, unlike gravitons which pass through it.

Brane: a membrane-like world in higher-dimensional space (called *bulk*) that can contain a universe.

Braneworld: Universe consisting of branes and bulk and all types of matter and forces.

graviton particles, and gravitons exist throughout the infinite bulk of *braneworld.* Gravity may have its own brane, where it is extremely powerful, but by the time it reaches our brane it has dissipated and is weak. Branes, like the strings described in string theory, can be *open* or *closed.* They resemble flexible membrane-like objects, something like a flexible plane *(an open brane)* or a flexible bubble *(a closed brane)*. Both an open or closed brane separate bulk. Open branes separate bulk like layers of cake, while closed branes trap a portion of bulk within the interior of its bubble. The *Randall-Sundrum modified model* has our brane embedded in an *infinite warped 5th*

dimension. As inhabitants on our brane we are unaware that all objects are warped in this 5th-dimension. Specifically, gravitons as they approach our brane are warped in the infinite warped 5th dimension   In other words, the gravitons are dispersed in this infinite expanse so that they lose their collective strength. The equations and established laws of physics still hold true with this model. The especially important new notion is the *infinite warped 5th dimension.* If the Randall-Sundrum model is true, it would resolve the *hierarchy problem,* which asks, *Why is gravity such a weak force?*.

mathematical knots may unleash computer power

Knots and computers are a seemingly unlikely duo even though at times computers freeze and seem to be tied up in knots. Today mathematicians and scientists are exploring a new breed of computers that are literally tied to knots.

Will mathematical braids, computers and quantum physics join forces?

Mathematically, a *knot* does not resemble the knots we normally think of such as square knots or slip knots, but is a knot with no loose ends. Yet, mathematical knots are real and appear in natural phenomena. In fact, in 1999 molecular biologists Titia de Lange and Jack Griffith discovered that normal chromosomes have no ends and hence resemble mathematical knots. Knots are also present in the twists and turns of DNA as they perform their biological functions.

Knot theory is also debuting in topological quantum field theory, and it's here *topological quantum computers* enter the picture. Conventional computers process information using *bits*—strings of ones and zeros indicating the "on" and "off" modes of electricity. Quantum computers, on the other hand, would theoretically be able to utilize the multiple energy states of a quantum(a unit of energy) called a *qubit*, processing not only strings of 0s and 1s but also a mixture of 0s and 1s simultaneously, thereby, performing multiple calculations with multiple answers all at the same time.

However, various conditions can influence the reliability of quantum computer information because subatomic particles can change unexpectedly and minute random disturbances in their environment (even radiation from light) can affect data encoded or manipulated by the quantum bits (qubits). These obstacles influence the stability of information that quantum computers will store and use, and this instability is referred to as *decoherence*. As a result, some computer scientists and mathematicians[1] are exploring the feasibility of the *topological quantum computer*. This quantum computer's approach is different from the qubit quantum computer.

It will not rely on single qubits to store and compute its data. Instead, the information will be stored and processed in a conglomeration of knots interwoven to form mathematical braids which will perform the quantum calculations by following the motion of *anyons*. *Anyons* are quasi-particles which appear and move in a 2-dimensional electron sheet of fluid which has been cooled to almost absolute zero while in a powerful magnetic field. Unlike electrons with -1 electron charge or protons with +1 charge, an anyon's charge can range from a fraction to an integer. As anyons move, they create knots which, in turn, form braids. The data will be "knotted up" i.e. stored in these braids of qubits which will keep the information more stable and less subject to decay. In qubit quantum computers, data would be stored on single particles, thereby, more easily impacting the qubit's stability. On the other hand, anyons are able to retain information of their knotty movements in the braids. All this will add up to more reliable data processing.

Imagine the number of problems nature solves everyday without our realizing it, be they biological problems involving our bodily functions, the molecular movements of atoms and particles, and all changes in our ever evolving universe. If we can tap into the powerful yet apparently innocuous properties of nature's "computers", we would be able to harness amazing tools. We have been able to put electricity to use with the design of today's microprocessor computers by "simply" figuring out how to describe mathematically the on/off states of electricity by utilizing a binary number system. Scientists are trying to design systems which can use the properties and energy states that nature holds in its atomic and subatomic levels. Perhaps it will be the system involving mathematical knots and braids that will harness the amazing properties of the quantum world. Regardless, in the final analysis, nature will be doing all the hard work which will continue to appear amazingly effortless.

The amount of information needed to be manipulated by a computer is enormous, especially when you consider it takes 8 bits to store a single letter of a word. *What's the difference between a bit and a qubit?* In conventional computers, a **bit** represents either 0 or 1, and things are communicated by stringing together 0s and 1s in order to write numbers and commands in the binary number system.

Qubits are what quantum computers will use for bits. A qubit can be a 0, a 1, or a mixture of 0s and 1s simultaneously. Thereby, quantum computers using qubits will be able to explore multiple paths of solutions at the same time because at the molecular level strange things take place. Here a particle can be in different places at the same moment in time. Yet in quantum mechanics, the moment a qubit is actually measured, it becomes a 0 or 1.

The unusual property of *how particles can be in more than one place at the same instant* is the signature trait of this minuscule world of particles.

---

[1] Alexei Kitaev, Zhenghan Wang, Edward Witten, Michael Freedman, John Preskill, Michael Larsen, and Seth Loyd are among the scientists and mathematicians exploring topological quantum computing.

multiverse theory & fractals

Not just one "Big" Bang?
—Not just one universe?
—Past, present, and future all exist simultaneously?
—At every point of the universe, a universe can or does exist?

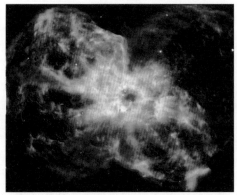

NASA's Hubble telescope captures the death of a star.

These are some of the ideas behind physicist Andrei Linde's *eternal chaotic inflationary* theory. Linde explains that *"Instead of a universe with a single law of physics, eternal chaotic inflation predicts a self-reproducing eternally existing multiverse where all possibilities can be realized."* [1]

The Big Bang theory leaves a lot of unanswered questions. Among these we have: What banged? What existed before the Big Bang? Why does our visible universe appear homogeneous and flat, even though general relativity explains space actually is curved?

Linde's theory points out that *"Our cosmic home grows, fluctuates and eternally reproduces itself in all possible forms, as if adjusting itself for all possible types of life that it can support."* [2] Eternal chaotic inflation explains there was no before. The past, present and future always existed. *"Each particular part of the [multiverse] may stem from a singularity somewhere in the past and it may end up in a singularity somewhere in the future."* [3] While some parts of the multiverse stop expanding and begin to contract, other universes are constantly being produced and inflating. There is no single beginning, as in the Big Bang, but of hosts of ever occurring "big" bangs. There is no before, the multiverse has always existed, and there is no after.

NASA's Spritzer telescope shows infant stars (reddish/pink dots) in cosmic clouds of gas and dust. The debris surrounding each has the potential of developing into planets.

Eternal chaotic inflationary theory explains why our universe appears flat and rather homogeneous. It all depends from what perspective you are viewing the multiverse. The observable universe which our instruments and experiments perceive is minute in comparison with the infinite array of universes in the multiverse. We happen to be living in a portion that appears flat and homogeneous. Thus far, this idea has been verified by NASA data experiments[4] which measured small variations in cosmic background radiation.

numbers and an algorithm(rule). The shapes and forms are infinite. Among the most famous is the Mandelbrot set—a fractal whose forms inhabit a universe of ever evolving universes. Its beauty and diversity

On the far left is a rendition of the Mandelbrot set. The three smaller images are zoomed in portions

This is just a peek at Linde's theory which encompasses many other ideas such as quantum fluctuations, the space-time continuum, parallel universes, universes expanding at exponential rates, dark energy, dark matter, collapsing and decaying universes.

Now consider the mathematical world of fractals—infinitely many geometric and random fractals evolving and self-replicating. All created by some set of objects or

of shapes is extraordinary. As one zooms in on any facet, new exciting forms appear which are in some way reminiscent of others. The same experience takes place as we zoom out. It's not one shape, not one design, not one universe, but infinite universes continually self-reproducing in an infinite expanse which is seemingly eternal, chaotic, and inflationary.

Physics and mathematics are always raising new questions and ideas and discovering new connections.

---

[1] *World Without End* by Scott Shackelford. *Stanford Magazine*. Nov./Dec, 2007
[2] Ibid.
[3] Ibid.
[4] Experiments conducted by NASA Cosmic Background Explorer satellite in 1990. In 2003 the WMAP(Wilkinson Microwave Anissotropy) satellite verified the prediction that our universe is flat.

mathematics & patterns

Patterns are a powerful concept, not merely as a model or template but as a repetition—repetitions as

Patterns in nature.

is 1 more or 1 less than a multiple of 2. The pattern observed in the Fibonacci sequence {1,1,2,3,5,8,...}, each term is the sum of the two previous terms, made it possible to describe it mathematically by $f_{n+1}=f_n + f_{n-1}$ where $f_1=1$ and $f_2=1$. Yet just because a mathematical expression can be found, does not automatically mean a pattern exists. For example, al-

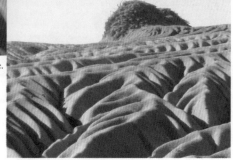

concrete as those found in the formation of the petals on a flower or as abstract as the recurring melody in a musical score. It's the repetition of a form or structure that makes a pattern such an invaluable mathematical tool—a tool which may harbor important hidden ideas or truths. That's why mathematicians have always been on the lookout for patterns. When a pattern surfaces, mathematicians try to describe it and predict its evolution with mathematical expressions or graphs. Some of the first mathematical patterns noticed appeared in simple numbers sequences, such as how the counting numbers increase by one unit at a time, how any even number can always be described as a multiple of 2 and any odd number

though mathematicians have developed various algorithms to generate π's endless decimal approximation, no pattern has been found among the billions of π's digits to accurately predict the next digit.[1] Similarly a mathematical expression for determining prime numbers has eluded mathematicians for centuries. Eratosthenes (275-194 B.C.) first developed a prime number sieve which determines prime numbers[2] up to an arbitrary number. Ever since then, mathematicians have been hunting for a simple way to find huge prime numbers. Even so, no simple equation or algorithm has yet been developed to sniff out primes.[3]

But finding patterns is not unique to mathematics. Probably the first patterns noticed by humans were those displayed by nature— patterns in the sand, patterns in the cycles of the waning and waxing of the moon, patterns of the seasons, patterns in the scales of a fish, patterns in crystals, patterns on shells, patterns in a honeycomb. Even objects that appear to be patternless such as thunderhead clouds or lava flow can be described with mathematics. Fractal geometry and computer technology have produced algorithms that are capable of reproducing such forms, which in the past escaped mathematical description. Add to these areas the mathematics of complexity, chaos theory, statistics, and fuzzy logic, and one wonders how phenomena still elude mathematics.

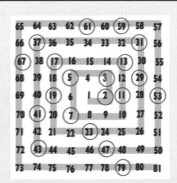

In 1963, mathematician Stanislaw Ulam noticed when he arranged the counting numbers in a spiral pattern, the primes tended to cluster. Some appeared along diagonal lines and others along horizontal and vertical lines. Using some of these clusters, mathematicians have developed mathematical expressions to predict subsequent primes.

This diagram shows how a pattern of dots can be interpreted mathematically. Each L-shaped dot pattern represents an odd number, yet all together they form a square, implying the sum of n odd counting numbers is n^2.

---

[1]Mathematicians Simon Plouffe, Fabrine Bellard and Jonathan Borwein developed an equation which lets one find any digit of π without calculating the previous digits—the catch is that the particular digit is in binary notation and, therefore, is either a 0 or a 1, which cannot be converted into its base ten digit.

[2]A prime number is a number that is only divisible by 1 and itself. Prime numbers are essential to cryptography, which continually becomes more and more important for securing secrets and protecting our privacy in today's digital world.

[3]In the summer of 2002, Mandindra Agrawal, Neeraj Kayal and Nitin Saxena of the Indian Institute of Technology (IIT) in Kanpur, India, put finishing touches on a new efficient method for finding large prime numbers. Their discovery took the mathematical world by surprise because it focused on a method that mathematicians had previously overlooked. The IIT team's work inspired other mathematicians who have been improving upon the method's speed for finding large primes.

pozzo's trompe l'oeil

Andrea Pozzo's famous trompe l'oeil on the ceiling of Sant'Ignatius in Rome, Italy.

Mastering the techniques of perspective, Italian artist Andrea Pozzo(1642-1709) wrote one of the first books on perspective, *Prospettiva di' pittori et architect*. His famous painting on the ceiling of the Church of Sant'Ignazio in Rome, Italy leaves you wondering if the figures are actually 3-dimensional with legs and arms appearing to come out of the ceiling as you view it at a spot marked by a disc on the floor of the nave. Today the church has a mirror placed at the disc's location to allow several people to simultaneously view the work. Although the ceiling is cylindrically shaped with a diameter of about 56 feet, Pozzo's major *trompe l'oeil* was to make the ceiling appear as if it were actually a dome. The story is that the church's neighbors did not want a dome to block their views.

hear no, see no, speak no
new mathematics

Mathematics leads in many directions and to many seemingly strange ideas. If the mathematics behind discoveries is consistent, the ideas are usually portals to new mathematical breakthroughs. Along the way there have always been mathematicians who would not accept or even consider the possibilities and usefulness of new math things that popped out of the expansion of mathematical concepts. Among these are:

• The Pythagoreans refused to consider √2 and other irrationals as numbers.

• Most mathematicians of the 16th & 17th century refused to accept negative numbers let alone imaginary, complex, quaternions, transcendental and transfinite numbers. During this era we have Nicholas Chuquet and Michael Stifel referring to negative numbers as absurd, and Girolamo Cardano giving negative numbers as solutions to equations even though he considered them as impossible answers. Blaise Pascal was reputed to have said *"I have known those who could not understand that to take four from zero there remains zero."* On the other hand, Albert Girard recognized complex numbers as formal solutions to equations which had no other solutions.

• From the late 1860s to the early 1900s traditional mathematicians refused to accept what they called mathematical anomalies/monsters that surfaced from the works of such mathematicians as Cantor, Wierstrass, Peano, Koch and Sierpinski, which led to today's evolution of fractal geometry.

$\sqrt{2}$ the first irrational number

-2 negative numbers

i imaginary numbers

$a + bi$ complex numbers

$\aleph_0$ transfinite numbers

π and e transcendental numbers

logic problems

Logic and logical thinking are at the core of mathematics. Here are two logic problems to sharpen your skills.

The Hat Problem

Three men are blindfolded. They are told that either a black or tan hat has been placed on each of their heads. The blindfolds are removed. The men see one another's hats. Then each, in turn, is asked "Do you see one or more black hats, and do you know what color hat you have on?" The first man answers *yes* to the first question and *no* to the second. The second man also answers *yes* and *no*. The third man answers *yes* and *no* . Then the first man is asked again if he knows what color he has on, and he says *yes*. How did he figure it out, and what color does he have on?

Which is which?

Each of the four cards is either black or red on one side, and each has either a square or a circle on the other side.

Which card(s) *must* you turn over to determine if every black card also has a square on its opposite side?

puzzlers

The **number-check puzzle**
This clever puzzle was the creation of the famous English puzzlist Henry Ernest Dudeney (1857-1930). On a  ring there are ten checks numbered 0 through 9. Divide them into three groups without removing them from the ring so that the product of the first and second group equals the third group of numbers. For example, if the groups were 632 and 890 and 7154, we get 632 x 890 ≠ 7154.

One of America's most famous puzzlist was Sam Loyd(1841-1911). When he was only 14 years old, he had his first puzzle published in a New York newspaper.

Friars' puzzle
Rearrange the 10 coins on the friars' board so that each row, column and diagonal has an even number of coins. Only one coin per square is allowed. Can you get 16 or more even lines ?

Can you make the Latin square?
A Latin square is a square formed by *n* rows and *n* columns of objects so that each entry appears only once in each row and column. Latin squares date back to 11th century when they were used as amulets in the Islam world. Leonhard Euler systematically studied Latin squares, and in 1779 posed the **thirty-six officers problem:** *There are 6 regiments of 6 officers of different rank. Is it possible to arrange 36 officers in a 6x6 formation so that no rank or regiment is repeated in any of the rows or columns?* Euler believed this problem had no solution. French mathematician Gaston

Tarry proved Euler's hypothesis in 1901. This and other such Latin square problems led to important discoveries in the field of combinatorics.

Can you rearrange the shapes in this 5x5 square so that they form a Latin square in which a shape appears no more than once in any of the square's rows, columns, or diagonals.

★	★	★	★	★
●	●	●	●	●
■	■	■	■	■
▲	▲	▲	▲	▲
⬢	⬢	⬢	⬢	⬢

famous mathematical goofs

Whether it's taking the wrong freeway exit, being late for an appointment, not balancing the check book correctly, or resolving a problem—we all make mistakes. Mistakes can be humbling and enlightening, and often insightful in unexpected ways.

Mathematics is no different. The following are some notable mathematical goofs.

A labor-of-love goof

Imagine doing 20 years of manual calculations, and thinking your work was perfect. That's what happened to William Shanks who believed he had correctly calculated π to 707 digits in 1873. But in 1944 D.F. Ferguson published a list of the first 620 decimals and discovered an error in Shank's 528th digit. Shanks used the formula—

$$\frac{\pi}{4} = 4\tan^{-1}\left(\tfrac{1}{5}\right) - \tan^{-1}\left(\tfrac{1}{239}\right)$$

developed by John Machin in 1706. Ferguson on the other hand used

$$\frac{\pi}{4} = 3\tan^{-1}\left(\tfrac{1}{4}\right) + \tan^{-1}\left(\tfrac{1}{20}\right) + \tan^{-1}\left(\tfrac{1}{1985}\right)$$

Considering how tedious the calculations by hand were, it is impressive that Shanks had it correct for 527 places!

π out to 707 places

3.14159265358979323846264338327950288419716939937510582097494459230781640628620899862803482534211706798214808651328230664709384460955058223172535940812848111745028410270193852110555964462294895493038196442881097566593344612847564823378678316527120190914564856692346034861045432664821339360726024914127372458700660631558817488152092096282925409171536436789259036001133053054882046652138414695194151160943305727036575959195309218611738193261179310511854807446237996274956735188575272489122793818301194912983367336244065664308602139494639522473719070217986094370277053921717629231767523846748184676694051320005681271452635608277857713427577896091736371787214684409012249534301 4...

Error of units

On December 11, 1998, the Mars Climate Orbiter was launched on its journey to Mars. This spacecraft was going to play an integral role in the exploration of the red planet. On September 23, 1999 a final rocket was fired to put the spacecraft in orbit around Mars. A mathematical error repeated with each of its rocket firings caused the craft to go astray. What caused the math error? Inconsistent units! The engineers failed to convert English pounds into metric newtons. One English pound of force equals 4.45 newtons. This difference caused the Mars Climate Orbiter to approach Mars at the wrong orbit, causing it to smash into the planet.

Not looking at the entire picture

In the 1930's, Edwin Hubble felt galaxies were evenly distributed. He proceeded to "prove" this by taking photos only in specific regions of space. His photos showed galaxies in roughly equal numbers. Scientists Harlow Shapely and Adelaide Ames took photos in the Northern Hemisphere's sky and found concentrations of galaxies. Their findings were confirmed by Clyde Tombaugh who found that galaxies are not evenly spread out but arranged in clusters and superclusters. Hubble's astronomical contributions were not overshadowed by this error. For his contributions to astronomy, in 1989 NASA named the Hubble Space Telescope after him.

The fatal blow

Mathematician Gottlob Frege had been working on the logical development of arithmetic. His first volume had been published and as the second and final volume was going to press he received a letter from Bertrand Russell concerning the *Russell paradox*[1]. The Russell paradox dealt a fatal blow to Frege's work. Realizing his mistake, Frege began the appendix of his second volume with this note *"A scientist can hardly encounter anything more undesirable than to have the foundation collapse just as his work is finished. I was put in this position by a letter from Bertrand Russell as the work was nearly through the press."*

Russell's paradox deals with the idea of membership of a set. A set is either a member of itself or not a member of itself. Refer to a set which does not contain itself as a member as *regular*. For example, the set of people does not include itself as a member, since it is not a person. Refer to a set which does contain itself as a member as *irregular*. An example is the set of sets with more than say five elements. Is the set of *all* regular sets regular or irregular? If it is regular, it cannot contain itself. But it is the set of all regular sets, thus it must contain all regular sets, namely itself. If it contains itself, it is irregular. If it is irregular, it contains itself as a member, but it is supposed to contain only regular sets.

something from nothing?

Over the years, infinity has caused and continues to cause all sorts of mathematical problems.

step 4: Solving this for S we get:
$$1 = S+S$$
$$1 = 2S$$
$$\frac{1}{2} = S$$

Does

$1-1+1-1+1-1+1-1+1-1+1-1+1-1+....=\frac{1}{2}$?

One interesting example is that which mathematician Gottfried Leibniz (1646-1716), one of the discoverers of differential and integral calculus, proposed about the sum of the series
$$1-1+1-1+1-1+1-1+1-1+1-1+.....$$
He contended it totaled 1/2.

Looking at and grouping this series in this manner:
$$(1-1)+(1-1)+(1-1)+(1-1)+(1-1)+...$$
seems to imply the series equals 0.

Here is how he arrived at 1/2.
He let **S** stand for the sum of
$$1-1+1-1+1-1+1-1+1-1+1-1+....$$
then reasoned as follows:

step 1
$$1-1+1-1+1-1+1-1+1-1+1-1+....=S.$$

step 2: Now group this as follows:
$$1-(1-1+1-1+1-1+1-1+1-1+1-1+...)=S$$

step 3: Now substitute S into *step 2:*
$$1-S = S$$

Does this mean that
$$(1-1)+(1-1)+(1-1)+(1-1)+(1-1)+...$$
$$=0+0+0+0+0+...= 1/2 ?$$
Beware: Playing with series which do not actually have sums, as this one, can lead to erroneous statements.

Mathematically speaking a *series* is the sum of a *sequence's terms.* For example, the sequence 1,2,3,4,5,6 can be associated with the finite series is 1+2+3+4+5+6 whose sum equals 21. Since a sequence can also be infinite, such as 1,2,3,4,5,6,... its series in this case can be written 1+2+3+4+5+6+... and its sum's value never stops increasing. Such series can create a lot of interesting problems. For example, which sequence has more terms 1+2+3+4+5+... or 1+3+5+7+8+...? Georg Cantor showed they actually have the same number of terms.

the fibonacci waltz

The Fibonacci sequence — 1,1,2,3,5,8,13,... — has been fascinating mathematicians and lay persons for centuries. The sequence happened to be the answer to an obscure problem Leonardo da Pisa, aka Fibonacci, proposed in his book *Liber Abaci*[1] which was published in 1202. But, it wasn't until the 19th century that Fibonacci's name became associated with the sequence. French mathematician Edouard Lucas attached Fibonacci's name to the sequence when the rabbit reproducing problem was included in a four volume set on recreational mathematics that Lucas was editing. Over the years the sequence has been discovered in many areas including nature, plants, the golden ratio, the equiangular spiral, the Pascal triangle, the golden rectangle, mathematics tricks, art and music. Some people contend that even Mozart may have subconsciously used the sequence in some of his compositions. Since each number of the Fibonacci sequence is connected to previous terms it's known as a recursive sequence. In fact each term of the Fibonacci sequence is produced by adding the two preceding numbers.

For an unusual and delightful treat and introduction to the Fibonacci sequence in music, listen to Ted Froberg's the *Fibonacci Waltz* by logging onto http://home.tampabay. rr.com/warhawks/FibonacciWaltz. html. He clearly explains how he simply related each Fibonacci number to the notes of the scale. Froberg then worked on defining its melody and rhythms to produce a lovely waltz which could theoretically go on indefinitely. You can also hear how the melody sounds with different scales.

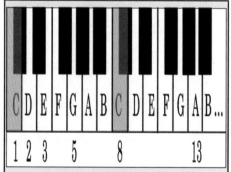

The keyboard above illustrates how the Fibonacci numbers are connected to the notes on the piano. To associate notes with subsequently larger Fibonacci numbers, Froberg ignores octaves, instead repeats the scale as shown in the gray shading. Thus the first six notes of the waltz would be:

C D E G C A ...

The hard work comes in deciding on the lengths of the notes, their rhyme, etc.

---

[1]*Liber Abaci* dealt with Hindu-Arabic numbers and computation with these numbers. The problem Fibonacci proposed involved a hypothetical group of rabbits and the numbers involved in their unique reproduction process.

quasi-periodic music and more

Although there are only a finite number of musical notes, when

The *Golden Bell Tower* solo exhibit by Akio Hizume in Hiroshima, Japan. June, 2007. Hizume explains how the exhibit illustrates three dimensions of quasi-periodic objects. The sculpture is the 3-D form, the shadows and layout are 2-D objects and finally the background music is 1-D.

signature, for example 3:4 time, can be thought of as its period. For traditional music scores this remains the same for every measure in the piece. For a 3:4 time signature every measure has 3 beats. On the other hand in *quasi-periodic music*, its time signature varies and is referred to as dynamic. The changes can be minuscule which at times may sound almost periodic, yet these small variations make it quasi-periodic.

Some composers like to experiment with quasi-periodic music and have linked it to mathematical patterns or sequences of numbers such as the Fibonacci numbers. Sculptor, architect, composer Akio Hizume has used these concepts to create *Fibonacci Kecak*. He explains that his computer program generates his music consisting of 9 periodic rhythmic parts which never repeat. Hizume points out that the rhythmic patterns change ever so slowly and never repeat.

these notes are combined with different rhythms and repetitions they can produce infinitely many combinations. In other words, there can be as many different musical pieces composed as there are different counting numbers. We are most familiar with musical pieces whose time signature remains consistent throughout the entire piece. A musical piece's time

Some musicians such as saxophonist Rudresh Mahanthapp and his quartet also use the Fibonacci sequence and aspects of number theory to jam their jazz. His *Further and In Between* uses the *cyclical number* 142857. The melodies are formed by linking these digits to semitones (half-step notes such as C and C# are a half-step apart on the scale). The number 142857 is a cyclical number with 6 digits and when it's multiplied by any digits between 1 and 6 results in a number having the same digits.

$$142857 \times 2 = 285714$$
$$142857 \times 3 = 428571$$
$$142857 \times 4 = 571428$$
$$142857 \times 5 = 714285$$

Mahanthapp's melodies of *The Decider* were created by mapping the Fibonacci sequence to the 12-tone musical scale. Mahanthapp claims that *"It sounds right no matter what key the others are compying[accompanying] in...I tried alternative sequences and they didn't have that property."*[1]

A number having n digits is called **cyclical** if when multiplied by any natural number less than n, the result contains the same digits as the original number.

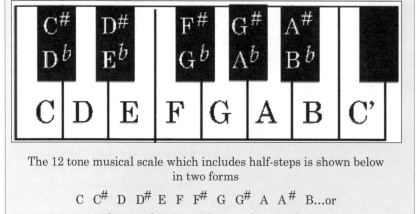

The 12 tone musical scale which includes half-steps is shown below in two forms

$$C \quad C^{\#} \quad D \quad D^{\#} \quad E \quad F \quad F^{\#} \quad G \quad G^{\#} \quad A \quad A^{\#} \quad B...or$$
$$C \quad D^{b} \quad D \quad E^{b} \quad E \quad F \quad G^{b} \quad G \quad A^{b} \quad A \quad B^{b} \quad B...$$

Fibonacci sequnece

$$1 \quad 1 \quad 2 \quad 3 \quad 5 \quad 8 \quad 13 \quad 21 \quad 34 \quad 55 \quad 89 \quad 144...$$

To hear clips of the musical pieces mentioned go to:
- **http://www.starcage.org/fibonaccikck.mid** for Aki Hizume
- **http://www.pirecordings.com/pi21/** for Rudresh Mahanthapp

---

[1] From *Riffs on Numerical Themes* by John Bohannon. Science vo. 315. January 26, 2007.

the mysterious antikythera mechanism

In 1900 the Antikythera mechanism was recovered from a shipwreck off the coast of the island

A major fragment of the Antikythera mechanism on display at the National Archeological Museum in Athens, Greece.

of Antikythera near Crete. Among the ancient artifacts of sculptures, chard, coins and amphoras was found a corroded device that was dated around 78 BCE. The device was first radiographed with gamma and x-rayed in 1971 by nuclear physicist Dr. C. Karkalos. The pictures revealed 2-D images of its internal gears and teeth. Subsequent radiographs were performed. Working from these images and the mathematical relationships that exist between toothed gear wheels, Derek de Solla Price of Yale University used the partial tooth counts to estimate the number of teeth on the gears and

made a schematic drawing for the mechanism's reconstruction. In 1974, Price suggested that the device was a calendrical sun and moon computing machine which was designed to tell time based on the movements of the Sun and Moon, the relationship of their eclipses and the movement of planets and others stars that were known during that period. *"The differential gear assembly used to take the difference between two rotations is the first found in the Antikythera mechanism, which is one of the most important achievements of ancient technology."[1]* Price was able to calculate that the large front gear had 235 teeth. 235 is a very significant number because it indicated that the device used the Metonic cycle, which was based on the Babylonian discovery that there are 235 lunar months in 19 years.[2] This concept was invaluable for keeping the solar and lunar calendars synchronized. A model was reconstructed from Price's investigation. He depicted it as consisting of pointers, dials, and more than thirty gears of various sized mesh on parallel levels with shafts rotating at different speeds. Unfortunately Price died before finishing his work on the mechanism. Although his model had some flaws it is the one on display at the National Archeological Museum in Athens. From his publication, *Gears from the Greeks: The Antikythera Mechanism, a Calendar Computer from Ca 80 B.C.,* Price wrote *"Nothing like this instrument is preserved elsewhere...from all we know about science and technology in the Hellenistic Age we should have felt that such a device could not exist.... [the mechanism] requires us to completely rethink our attitudes toward ancient Greek technology."* The original studies of its fragments proposed that the mechanism was a type of astrolabe used in seafaring, yet such technology did not appear in the Islamic countries until the 8th century and in Europe until the 12th century.

In 1990 stereo radiography using linear x-ray tomography was performed at the National Archeological Museum by a different group of researchers[3]. They reached a different opinion as to the function of the device, and hoped to make a new reconstruction that would illustrate a higher degree of functionality.

Then in 2005, the National Archeological Museum in Athens allowed a team of scientists using the X-Tek 3-D x-rays microfocusing computer tomography with the BladeRunner system and

On display at the National Archeological Museum in Athens is the first model of the Antikythera mechanism constructed by Derek de Solla Price, and donated to the museum. It continues to serve as a reference for future models.

Hewlett-Parkard's special imaging technology to scan the mechanism. These pictures clearly revealed the complex arrangements of the mechanism's 32 gears and plates. The process also showed the degree of corrosion of its material and that information proved helpful in its reconstruction. The mechanism was made of bronze, and its plates had inscriptions including the zodiac signs and months. The images were used to reproduce virtual images revealing how the mechanism would look without any corrosion. This imaging made the inscriptions far more visible and discernible revealing over 1,000 new letters and inscriptions for the researchers to study and translate. This more than doubled what had previously been seen. It took several months to complete the translations of all this information that appeared on the mechanism's 82 fragments. For example, on the

eclipses. Even a zodiac dial was identified on the front of the device.

The mechanism's dimensions are about 12" tall x 8" wide x 3-4" deep with about 30 hand-cut

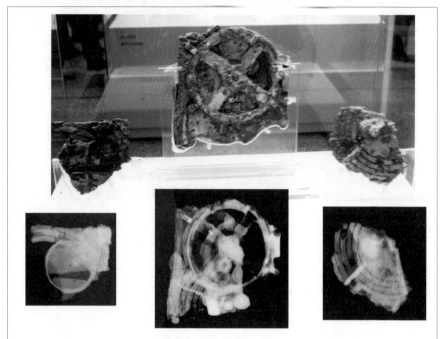

Three of the Antikythera fragments and their x-ray photos on display at the National Archeological Museum in Athens.

fragment labeled E were engraved the words "235 division on the spiral". Here a dial appears with the word "eliki" meaning spiral in Greek. This spiral dial's pointer had a small golden sphere representing the Sun. Equally impressive were the 235 distinct divisions which indicated the 235 hand cut teeth of one of its gears. On fragment #22, the number 233 is engraved indicating that this dial was based on the Saros cycle[4] and would point to the occurrence of

gears/wheels. Equally important is the special spiral rather than circular dials. New models of the device have been made. One such model illustrates the intricate movements the instrument performs. Turning its knob sets in motion its gears and dials showing the day of year pointer moving at a regular pace while the lunar and planetary pointers follow the orbits of celestial bodies which at times reverse directions following precisely the planets movements in the sky. The other dials on the back of the mechanism show how it also keeps track of the Saros and Metonic cycles. As research scientist Tom Malzbendera points out *"It is the first mechanical calculator known....Nothing as complex is known until you get to the Middle Ages, when people started building clocks."*[5] Who made this advanced mechanism "computer"? It may have been the Greek astronomer Hipparchus who lived on the island of Rhodes (circa 140-120 BCE), or Poseidonios who continued the school Hipparchus founded. Cicero makes three references to special mechanisms in his letters. In one letter he mentions Roman general Marcellus who coveted the booty of Archimedes' analog planetarium. In another letter he refers to the Rhodians who constructed such devices for exports to wealthy Romans, and in a third letter he writes of Poseidonios' device in which each revolution reproduces the daily motion of the sun, moon and the five planets. The most recent model for the Antikythera mechanism is that designed by AMRP(Antikythera Mechanism Research Project). Tony Freeth, mathematician at University of Cardiff, writes *"...in the AMRP model are a 223-month Saros eclipse prediction dial with an Exeligmos (Tripe Saros) dial as a subsidiary for the lower Back Dials and a complete reassessment of Price's Differential. We believe it is an epicyclic system that works in an entirely different way and realizes Hipparchos' theory of the Moon's irregular motion with a wonderfully ingenious design....In addition, the output of this system is to the front not the back dials... Mechanism [is]... an economic design with beautiful harmony and coherence...In addition all extant gearing(except a single unknown gear) can now be understood to be based on the two year cycles of the solar system from Babylonian astronomy: the Metonic Cycle and the Saros Cycle."*[6] For 3-dimensional animated views and additional information go to http://www.antiky theramechanism.org/.

---

[1] From the descriptive plaque on the Antikythera mechanism in the National Archeological Museum in Athens, Greece.

[2] This explains why a particular moon phase and its location repeats itself every 19 years. Further, the Greeks were able to more precisely synchronize the lunar and solar cycles by using their Callippic cycle, which was one day less than four Metonic cycles.

[3] Australian professor Al Bromley, English engineer M. Wright, Greek chemist E. Mangou.

[4] The Saros cycle was first used by Babylonian astronomers to predict the eclipses of the Sun and the Moon. They observed that 18 years, 11 days, and eight hours after an eclipse occurred a nearly identical eclipse will recur.

[5] *Secrets Unveiled* by Benjamin Pimented. San Francisco Chronicle, Nov. 30, 2006.

[6] *Old & New Models of the Antikythera Mechanism,* by Tony Freeth. The Antikythera Mechanism Research Project.

m-theory

M-theory, the inspiration of mathematical physicist Edward Witten of Princeton's Institute for Advanced Study, is one of the many Theory of Everything (TOE) to emerge. In 1994, Witten and Paul

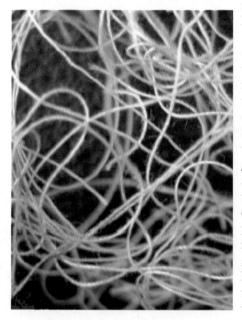

Townsend of Cambridge proposed that all the string/superstring theories (six in all) were actually part of the same TOE. That in fact, each of these string theories give distinct perspectives on TOE and are all parts of the single theory, which Witten called M-Theory.

The essence of M-theory and all string theories is the concept that

the building blocks of the universe are not atoms[1], molecules or particles, but 1-dimensional massless infinitesimal small strings which are 10^{-33}cm long (Planck length). Everything—humans, cats, rocks, electricity, heat, light—all forms of matter and energy (all forces)[2] is composed of strings. M-theory envisions the universe as 11-dimensional (10 spacial dimensions and 1 dimension of time) with its essence these infinitesimal vibrating strings. It's the strings' vibrations that distinguish them from one another. Each distinct vibration determines whether a string is a *boson* or a *fermion*. The bosons have an integer spin, and these are the particles which carry forces, e.g. a gluon carries strong nuclear force, a photon carries electromagnetic force, and a graviton carries gravitational force. The fermions have an odd half integer spin, and they are the particles which make up matter, including people, trees and even particles such as electrons and quarks.

Earlier string theories and super-string theories were formulated in 10 or 26 dimensions. The first string theory only dealt with boson particles, which meant it did not take matter into consideration. John Schwarz of the California In-stitute of Technology and Michael Green of Queen Mary's College in

London made a major breakthrough in 1984 when they introduced *supersymmetry* to string theory. This accounts for the prefix super in superstring theory, which considers both the matter and the energy that make up the universe. There are various versions of superstring theory—some only deal with closed strings, others only with open strings and some with both. The mathematical beauty of superstring theories is that they have no mathematical inconsistencies, unlike other TOEs. No infinite quantities or singularities appear in calculations involving high energy particle interactions, such as the interaction between two gravitons. In other words, the mathematics does not produce answers that are infinities[3]. Other attempts at finding a unified field theory for Quantum Field Theory(QFT) and General Relativity(GR) have produced mathematical results with infinities which contradicted their respective experimental results. But superstring theories are able to reconcile GR and QFT without creating these mathematical inconsistencies.

M-theory explains that at the moment of the Big Bang all its 11 dimensions were equal and encased in an infinitesimal small bubble or point. After the Big Bang, the universe split into a 4-dimensional bubble which began expanding while the remaining 7-dimensional part was compacted and tightly curled up in 7-dimensions.

In mathematics, an object has symmetry if it remains unchanged after it's undergone a transformation, such as a reflection about a line or rotation about a point. *Supersymmetry* is the symmetry that exists within particles and allows one particular particle to be transformed into another by rotating it until its spin matches the other. Particles of matter (fermions) can be transformed into particles of force (bosons) or vice versa. In essence, both types of particles are the same, just viewed from a different facet. When a fermion is transformed into a boson and then back to a fermion, the only change that has taken place is that the particle has moved in space—resulting in a transformation of space-time.

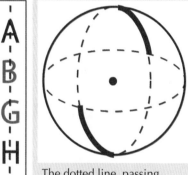

The dotted line passing through the letters shows which of these letters have *line symmetry*, that is the line cuts the red ones into identical halves while the green letters' halves do not match up. The sphere above illustrates *point symmetry*, in which every point of a sphere matches with another point on the sphere when it is reflected through the sphere's center.

Why all the dimensions? The answer goes back to 1919 when mathematician Theordr Kaluza wrote a short letter to Albert Einstein about Einstein's theory of gravity. In the note, he rewroto Einstein's mathematical work on the theory of gravity into 5-dimensions (4 spacial plus time) rather than Einstein's 4-dimensions (3 spacial plus time).

of gravity and Maxwell's electromagnetic force. It was along this vein that mathematicians/physicists introduced higher dimensions into string theories so that they could account for the various forces and forms of matter which exist. The various questions and problems that had appeared from previous traditional attempts at

One of several *star dust sculptures of the Eagle Nebula*. NASA explains that *"greater Eagle Nebula, M16, is actually a giant evaporating shell of gas and dust inside of which is a growing cavity filled with a spectacular stellar nursery currently forming an open cluster of stars."* M-theory explains that all matter and energy is determined by minuscule vibrating strings harbored within the eleven dimension of our universe. Photo taken 12/9/2007 courtesy of the Hubble Heritage Team, (STScI/AURA), ESA, NASA.

To Einstein's surprise, not only was this elegant mathematical work consistent (had no mathematical singularities or infinities), but the 5th dimension of gravity reproduced mathematician/theoretical physicist James Maxwell's[4] electromagnetic force. In other words, translating Einstein's work into a higher dimension unified Einstein's theory

unifying QFT and GR disappeared when the work was calculated in higher dimensions.

What does a string look like? A 1-dimensional string can resemble a minuscule open or closed (loop) piece of thread. By adding the 11th dimension to M-theory, a 2-dimensional minuscule tube-like string, similar to a flexible straw, can also be included. These are called membranes or branes. A 1-dimensional string is now called a 1-brane and the 2-D membrane a 2-brane. But that does not preclude 0-brane strings or 3, 4...brane strings. All the strings (branes) with their vibrations will help account for the menagerie of particles and forces, known and unknown. Strings as actual membranes and their manifold universes are ideas currently being delved into in M-theory

In addition to possessing supersymmetry, M-theory also has a new symmetry called *duality* in which two objects can change places with one another. For example, suppose the radius of our universe is R and that of the 7-dimensional compacted universe is 1/R. As our universe expands, the radius R gets larger and larger while 1/R gets smaller and smaller. Duality explains that R and 1/R are interchangeable and one's change

effects the other, thereby making it possible for the universe to expand and contract simultaneously. Many many aspects of M-theory remain a *mystery*. The mathematics fits together, but as yet so much has not been verified by actual experiments. Perhaps as new mathematics and experiments are designed pieces to the puzzle will fit together. It seems as one piece falls into place, a new problem or idea surfaces. What new innovations and discoveries will M-theory lead to? The one thing certain is that M-theory and other TOEs churn the imagination and boggle the mind.

---

[1] An atom consists of electrons and a nucleus. When the nucleus was split, it was discovered to consist of other particles, such as quarks. These particles and the forces behind them are believed to be vibrating strings. In 1964, Murray Gell-Mann discovered the quark by atom spitting.

[2] The four elementary forces are electromagnetic force; weak force (radioactive decay); strong force, which holds the nucleus together; and gravitation force (gravity).

[3] For example, if you were working on a problem that resulted in 2/0—the answer is undefined. It's an infinity.

[4] James Maxwell's work touch many scientific fields and areas. His ideas were major stepping stones to special relativity theory and quatum theory, and Einstein's famous equation relatng energy and mass, $E=mc^2$, was an offshoot of Maxwell's work.

dissections—cutting up mathematics

Dissection problems can be amusing, challenging and enlightening. Here are two to ponder.

Impossible cut-up?
Get a rectangular sheet of paper or an index card. Using no tape or glue, cut, fold and transform the rectangle into this mystery shape.

Dudeney's diabolical dissection problem

H.E.Dudeney was a master puzzle maker. Here we are asked to cut this figure into two pieces which can fit together to form a square.

(1) First try it ignoring the 4 holes.

(2) His diabolic challenge is to cut the figure into two pieces that form a square without cutting through any of the four holes.

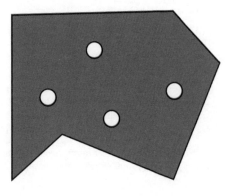

Solutions are given in the solution section at the back of the book.

In April of 1829 Niels Henrik Abel died in Froland, Norway. Not yet 27 years old, his short life nevertheless was to have a major influence on the progress of mathematics.

Bernt Holmboë when Abel was in his mid-teens. Seeing his student's talent, Holmboë not only encouraged and guided Abel's work but also helped him raise money for his family's support by finding him tu-

quintic equations
elliptic functions
Abelian integrals
Abelian equations
abelian groups
Abelian functions
Abel's theorem

This stamp depicting Abel was issued by Norway in 2002 on the 200 year anniversary of his birth. Based on a painting by Johan Gorbitz in 1826, it is the only known portrait of Abel.

Born in the village of Findö, Norway in August of 1802, Abel was the eldest of seven children. His father, a pastor, died when Abel was 18 and the responsibility of his family's support was now his. Suddenly finding himself the primary bread winner of the household made life difficult and trying for Abel, yet he always maintained a positive outlook. His passion for mathematics and his desire to learn were his inspiration.

His mathematical abilities were first recognized by his teacher,

toring jobs and sponsors. Abel made his first major mathematical discovery at the age of 18. Many of us are familiar with the quadratic formula—the formula used to solve any second degree equation by simply plugging in its coefficients. Similar methods had been found in the 1500s for solving cubic and forth degree equations, but a formula for solving a *5th degree equation* (quintic) had eluded mathematicians for at least another two centuries. In 1820, Abel felt he had found a general way to solve any quintic equation. Holmboë sent Abel's solution to be reviewed by a mathematician friend, who fortunately asked Abel to elaborate on some points. Upon reviewing his work, Abel found an error in his work, which led him to conclude and prove that no such formula (i.e. using radicals) was possible for the quintic or for any higher degree equations. Abel published a memoir of it as his major accomplishment. In fact, he felt so keenly about the work that he sent a copy to Carl Gauss for review. For whatever reason, Gauss never bothered to look at it.

Regardless, Abel was not discouraged and was convinced his work was of significant importance. Desiring to pursue further studies and research Abel felt he needed to go to France and Germany to consult with other mathematicians on topics of interest. He was finally given a very small stipend from the Norwegian government for two years abroad. This, along with his continual help and encouragement from Holmboë, enabled him to leave. His first stop was Berlin, where he met August Crelle. Crelle was about to publish the first purely mathematical periodical, which focused on new/current mathematical ideas. Immediately recognizing the importance and quality of Abel's work, Crelle highlighted Abel's work in the first three editions of the magazine (*Journal für die reine und angewandte Mathematik*), including a revised and updated version of his work on the quintic. Crelle tried to persuade Abel to stay

in Berlin, and apply for a position at the University of Berlin, but Abel felt he needed to continue his travels to France. In Paris in 1826, he hoped to meet with such famous mathematicians as Cauchy, Legendre, Poisson, Fourier, Ampere, Dirichlet, Lacroix—to discuss mathematical ideas, but unfortunately his contact with them was merely superficial. While in Paris he was able finish work on his now famous Abel theorem dealing with transcendental functions. Abel felt he was fortunate to get his work to Augustin Cauchy (1789-1857) who at the time agreed to present it to the prestigious French Academy of Science, but Cauchy hardly took time to look at Abel's work. Apparently Cauchy was too caught up in his own work and had another colleague present Abel's paper. The Academy then appointed Legendre and Cauchy to referee the work. Legendre, 74 at the time and with poor eyesight, found the copy of Abel's paper too faint and felt a clearer and neater copy should be made. Cauchy then took the paper with him, and eventually forgot about it and apparently lost it. Abel was disappointed and disillusioned when nothing became of his paper. At this point his finances and his health were declining, and he left Paris after being diagnosed with tuberculosis. In Norway he continued to work on his mathematics and submitted articles to Crelle's journal, but his health deteriorated quickly. His final days were spent at the home of an English couple who lived in Norway. There he was cared for by his fiancé, who was a governess for the English couple. Two days after his death a letter came to Abel from Crelle with news that the University of Berlin had appointed Abel a professor of mathematics.

What became of Abel's monumental work? If it had not been for a letter by German mathematician Carl Jacobi (working on a topic similar to Abel's) to the Academy of Sciences in March of 1829, would Abel's work ever have been uncovered? Jacobi wrote asking about Abel's discovery, whether anyone had seen Abel's work, and what had become of this work since it had been submitted two years earlier to the Academy. The lost manuscript finally became a diplomatic affair, with the Norwegian consul in Paris intervening. The Academy of Sciences pressed Cauchy, and Abel's lost paper finally surfaced in 1830. As a result the Academy of Sciences awarded the Grand Prize in Mathematics jointly to Jacobi and Abel. Tragically late for Abel .

Abel's work spawned various ideas and work. Included among them are: abelian groups. elliptic functions, theory of equations with such names Abelian integrals, Abelian equations, abelian groups, Abelian functions. In 2002 the Norwegian government decided to recognize work and excellence in mathematics by establishing the Abel Prize Fund and appointed a committee from the Norwegian Academy of Science and Letters to establish the Abel Prize.

In 2002 Norway issued a coin commemorating Abel's 200th birthday.

The establishment of the prize commemorated the 200th anniversary of the birth of Niels Hendrick Abel, and thereby recognized the ingenious Norwegian mathematician. The Abel Prize makes an annual award of $875,000 and is on a par with the Nobel Prize, which offers no award in the area of mathematics. The first Abel Prize[1] was given in 2003 to French mathematician Jean-Pierre Serre for his role in shaping topology, algebraic geometry and number theory.

---

[1]Subsequent Abel Prizes: **2004**: Michael F. Atiyah and Isadore M. Singer for their discovery and proof of the index theorem, bringing together topology, geometry and analysis, and their work bridging mathematics and theoretical physics. **2005**: Peter D. Lax for his work on the theory and application of partial differential equations. **2006**: Lennart Carleson for his work in harmonic analysis and the theory of smooth dynamical systems. **2007**: S. R. Srinivasa Varadhan for his work in probability theory and a unified theory of large deviation.

nanotechnology—then and now

The origin of nanotechnology dates back to 1959 when Richard Feynman, Nobel Prize recipient for fundamental work on quantum electrodynamics, introduced the scientific community to the concept of

Fine motion controller for molecular manipulations ©Institute for Molecular Manufacturing. www.imm.org. Courtesy of Foresight Institute.

molecular manufacturing in his historic talk, *There's Plenty of Room at the Bottom*. He stated *"the principles of physics, as far as I can see, do not speak against the possibility of maneuvering things atom by atom...a development which I think cannot be avoided."* He spoke of machines designed to make ever smaller machines, again and again. At the time, his talk did not get much attention, but today it is considered a milestone in the history of nanotechnology.

But the actual field of nanotechnology was not launched until 1986 when Eric Drexler, who had coined the term nanotechnology, wrote the book *Engines of Creation: The Coming Era of Nanotechnology* and an article *Molecular Engineering: An Approach to the Development of General Capabilities for Molecular Manipulations* (which first appeared in the 1981 Proceedings of National Academy of Sciences). Drexler described nano-machines having engineered precision and most importantly he introduced the idea of nano-assemblers, self replicating machines built from the bottom up atom by atom. Drexler also spoke of the possibility of *nano-catastrophes,* maverick machines getting out of hand and taking over everything on Earth, reducing it to gray goo.

Drexler's work and ideas engaged the imaginations of many scientists, and questions of the feasibility of self-replicating nanomachines arose. In particular, chemist Richard Smalley batted around ideas with Drexler. Smalley is famous for his discovery of buckyballs and carbon nanotubes for which he shared a Nobel Prize in chemistry. Smalley concluded that Drexler's concept of nanotechnology involving self-replicating assemblers was impossi-

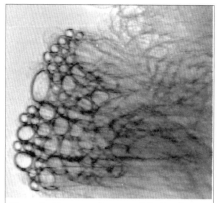

Nanotube-like ropes. Courtesy of the Smalley Research Group, Rice University, TX.

ty on nanotechnology. In 1992 Drexler had testified before the Senate Subcommittee on Science, Technologies and Space at a hearing on new technologies for sustaining the world. He described nanotechnology as the science of the future and advocated government research and development funding. The same year he published *Nanosystems: Molecular Machinery, manufacturing and Computation.*

Today some scientists are calling Drexler's molecular manufacturing unrealistic, and some scientists fear that because Drexler paints both the good and bad sides of nanotechnology it may negatively impact their research and funding.

ble, and in 1999 he testified to a Congressional subcommittee claiming Drexler's nanotechnology *"will always remain a fantasy ...there are simple facts of nature that prevent it from becoming a reality."* In addition, Smalley feared that Drexler's ideas would dampen the general public's attitude toward nanotechnology.

Until Smalley came on the scene, Drexler was considered the authori-

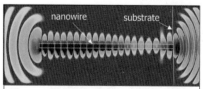

The complex interaction between light and nanometer structures, like wires, has possibilities as new technology for such devices as nanolasers. National Security Council(NAS) researchers are studying the possiblity of nanolasers in which light emission from a semiconductor nanowire 10-100 nanometers wide and a few micrometers long would function as a laser. Lasers made from arrays of these wires have many potential applications in communications and sensing for NASA. Illustration courtesy of NASA Ames Research Center.

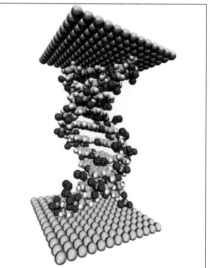

This example of an engineered DNA strand between metal atom contacts could function as a molecular electronics device. Such molecules and nanostructures are expected to revolutionize electronics, and may be particularly useful when electrical power is limited. Illustration courtesy of NASA Ames Research Center.

beware of nano-hazards

As with most scientific break-throughs the excitement of a new discovery often overshadows possibilities of potential scientific disasters. For example, splitting the atom produced powerful new energy sources, but along with it came atomic bombs, the potential of atomic plant meltdowns and the devastation of radioactivity.

Perhaps in hopes of sidestepping negative attitudes, Richard Smalley pushed for redefining nanotechnology. Instead of viewing it as molecular manufacturing with self-replicating nanobots in assembly line fashion working at exponential rates, he broadened the definition of nanotechnology by describing it as any technology that deals with work and products at the atomic level specifically focusing on nano-scale products rather than nanomachines. Politics and economics have taken nanotechnology in this new direction. The notion of nanotechnology products has attracted many businesses and scientists. Smalley felt his approach would be more public friendly by focusing on prod-

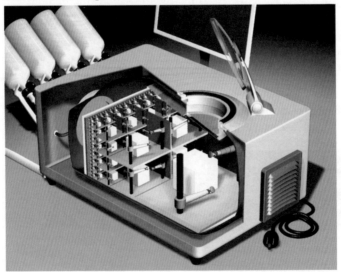

Eric Drexler's Nanotechnology Desktop Factory.[1] ©Institute for Molecular Manufacturing. www.imm.org. Courtesy of Foresight Institute. Image by John Burch. Lizard Fire Studios. http://www.lizardfire.com.

As described by the Foresight Institute: *"This machine represents a safe and practical approach to nanotechnology production. Raw materials enter on the left and finished products are removed through the top port under computer control. While this machine can produce objects roughly a decimeter on a side, the first and less expensive versions of this machine will be small and only able to make microscopic objects. However, medical sensors and body repair nanobots are also tiny and therefore most appropriate for production in such machines."*

ucts and not emphasizing potential nanotechnology dangers and possible disasters. In 2001 the US government officially established NNI (National Nanotech Initiative) focusing on Smalley's vision of nanotechnology products.

Meanwhile Drexler, the first US scientist to receive a PhD in nanotechnology (MIT 1991), now found himself out of the nanotechnology loop. In 1986 Drexler and Christine Peterson had cofounded the nonprofit organization Foresight Institute with emphasis on the importance of responsible nanotechnology science. The Foresight Institute website posted a set of guidelines in 1999 to promote safe and responsible nanotechnology work. Drexler hoped these safeguards for self-regulation would be embraced by nanocompanies and nanoscientists in order to avoid nanotech disasters. Thus far no nanotech businesses have signed onto the Foresight Guidelines. The Foresight Guidelines on Molecular Nanotechnology state that the Guidelines *"might eventually be enforced via a variety of means, possibly including lab certifications, randomized open inspections, professional society guidelines and peer pressure, insurance requirements and policies, stiff legal and economic penalties for violations, and other sanctions....Rather than focus on scenarios of runaway replicators, we should anticipate how molecular manufacturing can be used to improve our health and quality of life, restore the environ-*

ment, and prevent acts of aggression."

In the meantime, Drexler also revised his idea of self-replicating assemblers, saying that it is both safer and feasible to create *desktop nanofactories.* These factories would have various ports for robotic machines to carry out the assembling process along a conveyor belt. The robotic appendages would not be able to function independently of the nanoplant. When one appendage is removed its ability to work ceases, even as a light bulb ceases to function when removed from a lamp.

A **nano-something** is one-billionth of something, e.g. nanometer is 1/1,000,000,000 meter. Nanotechnologists work with objects between the size of 0.1 to 100 nanometers. The human eye can see things around the size of 10,000 nanometers across.

Where does mathematics appear in nanotechnology? It appears just about everywhere. It provides the numbers to measure and describe nanostuff, produces virtual animation on how such mechanisms would function, does calculations and provides equations describing behavior of nanomachines, looks to knot theory to explore the shapes which nano stuff can assume, and uses complexity and chaos theories to explain possible actions and problems of nanounits. Mathematics has a very big role in the very small world of nanotechnology. Without mathematics it would be futile to even consider delving into it.

nanotechnology watch dog?

Products evolving from Richard Smalley's definition of nanotechnology include such things as nanowires for electronic and optical devices, nanofibrils for hearing devices, wrinkle creams that penetrate deeply into the skin, super strong nanocables (many times stronger than steel) for body armor

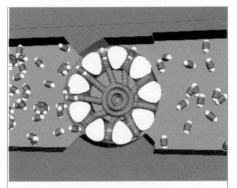

A design of a nano-rotor which is designed to purify feedstock. The left side would select acetylene molecules and reject larger molecules in the purification process. As it rotates the selected molecules are pushed out on the right into a bath of pure feedstock. ©Institute for Molecular Manufacturing. www.imm.org. Courtesy of Foresight Institute.

and elevator cables, nanofabrics which resist wrinkles and dirt. The possibilities seem endless.[1] But, hold on! **Presently there are no nanoproduct controls in place.** Government funding offers no guidelines or watchdog agency. In fact, in 2004, only 11% of government monies earmarked for nanotechnology were to be used for health and environmental studies. *"Nanoparticles can be ingested, in-*

haled, and absorbed through the skin. They also appear to cross the blood-brain barrier, nature's adaptation for blocking foreign substances in the bloodstream from reaching and disrupting the central nervous system."[2] On July 25th 2007 FDA Task Force report on Nanotechnolgy, the FDA maintains that regardless of the 'special properties' of nanomaterials, their use does not need to be regulated nor do products containing nanomaterial need to be labeled as such. Smalley, who himself has a biotechnology business involving nanotubes, is seeking to assuage public fears that arose from the possibility of nanoglitches alluded to in Drexler's vision of nanotechnology.

Studies show examples in which nanoproduct based technologies have not been carefully monitored and have resulted in unsafe practices. For example, toxic pathologist Vyvyan Howard at the University of Liverpool, England reviewed past studies and found that toxicity of particles come predominantly from its nanosize rather than from the material used for manufacturing. A study from Southern Methodist University in Dallas found that largemouth bass exposed to fullerene buckyballs in water for 48 hours developed brain damage.

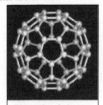

C60 molecule is an example of a fullerene (a caged molecule in the shape of a polyhedron composed of carbon atoms. Courtesy of Richaivo Saito, Japan.

Even though no current products contain buckyballs they do contain carbon nanotubes which on exposed rats penetrated cells and resulted in lung abnormalities. In the February 2004 edition of *The Ecologist* Jim Thomas points out that the Food and Drug Administration is *"privately admitting they have made a mistake in letting nanoproducts onto the market without safety studies, and are looking for ways to tweak existing regulations."*

The excitement and potential of new and incredible innovations often overshadow possible negative aspects. For example, most recently genetic engineering produced genetically modified food and plants. In some instances these changes helped overcome plant diseases and shortages, and in other cases they negatively impacted the ecosystem and affected human allergies. Ill effects might outweigh the benefits if studies and science rush to produce products before complete and thorough analyses of cause and effect have been carried out. Nanotechnology will be the manufacturing and technology revolution of the future, but its success and positive impact depends on our priorities. Will the nano revolution be fueled by economics driven by profits or be guided by seeking improvements in the quality of life?

Technically anything between the size of of 0.1 and 100 nanometers can be considered a nano-object. Many nanoproducts geared for everyday consumers are hitting the market. Here we find:

• *nano-glaze,* called ActiveCare, manufactured by Villeroy & Boch, is made with silver ions and used in basins to inhibit the growth of fungus and bacteria.

• *nano-water and fog repellents* called Clarity Defender and Fog Eliminator are produced by Nanofilms and used on windshields, glasses, goggles. The products' molecules self-assemble into a fine 10-nanometers thick coating on the surface which repels water and moisture.

• *nano-facial cream,* Anti-Aging Formula by Leunesse, is able to work its way down into layers of skin because its minute molecular size is 400 nanometers. It is technically too large to qualify as a nanoproduct.

• *nano-fabric guard,* Nano-Care made by Nano-Tex, resists staining and repels liquid.

• *nano-golf balls* made by Nano-Dynamics are supposed to correct the wobble and drift of a struck golf ball.

• *nanotubes in a set of bike components* claimed to be lighter and stronger than traditional parts are used by Easton Sports.

For other nanoproducts and breakthroughs check the Internet and websites such as scientifica.com.

[1] The US government website, http://www.nano.gov, reports that *"Some other current uses that are already in the marketplace include:* • *Burn and wound dressings* • *Water filtration* • *Catalysis* • *A dental-bonding agent* • *Step assists on vans.* • *Coatings for easier cleaning glass* • *Bumpers and catalytic converters on cars* • *Protective and glare-reducing coatings for eyeglasses and cars* • *Sunscreens and cosmetics.* • *Longer-lasting tennis balls.* • *Light-weight, stronger tennis racquets.* • *Stain-free clothing and mattresses.* • *Ink."*

[2] *Nanotech Under the Microscope* by Anne Geske, UTNE, July-August 2004.

balancing problems

The Babylonians were probably the first to develop weights to measure objects, rather than just compare the weights of two objects. They even developed different size stones to weigh different commodities.

Problems using balancing scales and weights were very popular during the 1600s, when there were only a few types of scales for measuring weights.

Step back to the 15th century and tackle these two balance scale problems.

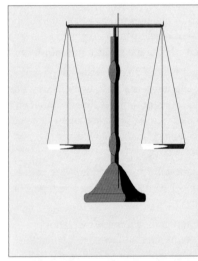

Problem (1) is credited to Claude Gaspar Bachet(1581-1638).

Problem (2)'s creator is unknown.

(1) Using a balance scale, what is the smallest number of weights needed to determine the weight of any object with an integral weight between 1 and 40.

(2) Using a balance scale, suppose you have 12 coins, and one of these is counterfeit (it is either lighter or heavier). In just three weighings, determine which is the counterfeit coin and if it is lighter or heavier.

Which is for real?

Suppose you have 5 stacks with 20 supposedly gold coins in each stack. Each authentic gold coin weighs 10 grams, but two of the stacks are composed of only counterfeit coins weighing 11 grams each. You are given a scale that weighs grams. Figure out a way to determine the counterfeit stacks in one weighing using this scale.

out, out damn nine

The way to check addition or multiplication work by *casting out nines* first appeared in the work of 9th century Arab mathematicians. The method is tied to the *digital root* of a number. Every number has a unique *digital root*. For example, the number 235 has digital root 1 because 2+3+5=**10**=1+0=**1**. And the number 199,999 also ends up having the digital root **1** because 1+9+9+9+9+9 =**1**. In fact, any 9s appearing in a number do not effect the outcome of that number's digital root because any sum of 9s ends up as a 9, and the digital root of 9n (where n is any digit) comes out to be n. So any 9 digits of a number can be *cast out* before adding up the rest of the digits to find its digital root.

The process of *casting out nines* and finding a number's *digital root* has two properties that lets you check addition or multiplication problems.

Here's how it works: Letting the symbol d(n) stands for the digital root of a number, **Property 1** states
    d(a) +d(b) =d(d(a)+d(b)).
This means the sum of the each digital root of the sum of two numbers is the same as the digital root of the sum. So the digital root of each number in a sum is determined and checked against the digital root of the problem's sum. Suppose

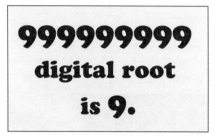

999999999
digital root
is 9.

we *cast out of nines* to check if 3,491+645 is 4126, and find the digital roots do not match: 8+6=5 ≠ 4, indicating a mistake has been made.

The same process works for multiplication, **Property 2:**
    d(a) · d(b)=d(d(a) · (d(b))
Casting out nines to check 57·403 = 22,071 we get 3·1=3 = 3. Thus, this problem seems to be correct, *but casting out nines does not work for all types of mistakes.* It will not work if you accidentally interchange digits of a number **or** if you replaced a 9 by a 0 because these two processes do not affect the digital root. In the above problem, the answer should have been 22,971, rather than 22,071, a mistake was made because the 9 was replaced by 0 and thus did not effect the result of its digital root.

Even though the digital roots for the problem 83+56≠193 match, the digits in the supposed sum 193 were mistakenly interchanged. It should have read 139.

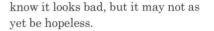

global warming & the math behind it

The data is in, and the Earth flunked its physical. Over the past 10,000 years the Earth has been in relatively good health. It has been

This image of Hurricane Katrina was taken August 27,2005 by NASA/ Jeff Schmaltz, MODIS(the Moderate Resolution Imaging Spectroradiometer)Lands Rapid Response Team.

know it looks bad, but it may not as yet be hopeless.

The Earth's climate is an intricate system with so many known and unknown factors that even advanced computer models with inputted data from satellites and other weather readings still cannot always make predictions accurately. The models are improving, but it is impossible to account for everything in spite of the sophistication of our computers and programs. Mathematically speaking, climate is a *non-linear complex system* in which infinite factors come into play. Minute changes in initial conditions can have monumental results in weather. This is the famous *butterfly effect* at work. In long term forecasts mathematical models are more successful in predicting climate change than in short term weather forecasts. Science has made tremendous headway in uncovering and recording what things affect climate change. Today the focus of these models is on *global warming*.

able to support its biological systems and where disasters such as plagues, earthquakes, floods, volcanoes occurred, it's been able to restore the health in those regions. Since the industrial age, however, many adverse human activities have been taking their toll on many of the Earth's life forms including humans themselves. The prognosis for its ability to sustain life as we

Global warming is the increase in the Earth's surface temperature due to the build up of greenhouse gases, especially carbon dioxide, released into the *atmosphere*. Carbon dioxide levels resulting from natural occurrences remained between 180 to 300ppm(**parts per million**) over the past 650,000 years. Industrialization has increased the carbon dioxide levels to 380 ppm which is about 25% higher than the pre-industrial era.

In 1824 mathematician Joseph Fourier discovered the ***greenhouse effect***, namely, how the atmosphere warms the Earth. The ***atmosphere*** is the layer of air that surrounds the Earth and reaches over 560 km from the Earth's surface. It shields the Earth's surface from radiation and the cold vacuum of space. The atmosphere recycles water and chemicals, absorbs the Sun's energy while working with electrical and magnetic forces to produce a mild climate. The atmosphere is approximately 78% nitrogen, 21% oxygen, 1% argon, 7% water vapor, and many other components including greenhouse gases. It consists of the troposphere, stratosphere, mesosphere and the thermosphere.

Today we know that climate is a complex system affected by an infinite number of constantly changing factors. Among these are some 30 greenhouse gases of which carbon dioxide, water vapor, methane, nitrous oxide, hydrochloroflourides (synthetic gases) are just a few, but comprise the largest share. Each of the greenhouse gases has a different half-life which determines how long it remains in the atmosphere. Greenhouse gases have an important role in maintaining a liveable average temperature over the Earth's surface in which diverse life forms thrive in various ecosystems. The atmosphere's gases create a greenhouse effect that insulates, regulates, and maintains the Earth's average surface temperature at around 58°F. If it were not for this greenhouse effect the Earth's surface temperature would average about 0°F.

Carbon dioxide is a *natural* byproduct of respiration and other functions of plants, animals and the environment. Since the industrial age carbon dioxide has also become a byproduct of such human activities as manufacturing, farming, clearing lands, and energy consumption generated by use of fossil fuels such as oil and coal. The increase in population[1] and the constant increase in the use of fuels and the rise of energy consumption account for the increased levels of carbon dioxide. *Why are we concentrating our attention on carbon dioxide?* Water vapor and carbon dioxide are the largest contributors to greenhouse warming. Water vapor is not considered an anthropogenic gas (i.e. humans do not directly increase water vapor levels even though they do so indirectly by their production of other greenhouse gases). The atmospheric concentrations of other greenhouse gases such as nitrous oxide, methane, ozone, nitrous oxide and sulfur hexaflouride are very low compared to carbon dioxide, thereby making carbon dioxide's greenhouse effect far greater even though many of these other gases trap heat more

especially carbon dioxide. Some of the gases' molecules remain in the atmosphere while others escape into space. The atmosphere's carbon dioxide level is increasing dramatically and constantly making the Earth's surface temperature rise.[2]

Between February and March 2002, Larsen B ice shelf collapsed and broke away from the Antarctic Peninsula. Photo by NASA/GSFC/LaRC/JPL, MISR Team

efficiently than carbon dioxide. As carbon dioxide and the other greenhouse gases are produced many of their molecules remain in the atmosphere for the duration of their half-life. In carbon dioxide's case that is 100 years. Heat generated by the Sun is trapped by the atmosphere's greenhouse gases thereby creating more gases,

We now know that it is not a question of whether global warming is taking place, but predicting when it will reach catastrophic levels and what measures we can take to stem its progression. We know about the atmospheric pollutants, especially those which humans have contributed. We know about which pollutants affect the Earth's ozone layer. We know about acid rain and smog. We know that global warming is not science fiction. Unless we take immediate steps directed by scientific findings, we will reach a point where the ongoing devastating cycles cannot be stemmed let alone reversed. Mathematical models point to a global warming tipping point. The Bush administration was in denial about the health of the Earth. Ultimately it modified its stance, but valuable time was lost. We acknowledge and are grateful for the ongoing work of Al Gore and many other individuals whose efforts have awakened us to the challenge. Each of us can do something to modify our use of energy from changing the type of light bulbs we use, carpooling, driving less and more slowly to using green energy/alternative renewable energies, to urging/ demanding local, state and Federal governments involvement[3]. Is our inertia and present comfort worth gambling the health and life of our planet?

Non-linear complex systems.

A complex system is composed of infinitely many interdependent changing facets that play off of one another. Examples include weather, the body's physiology, the brain, a nation's economy, a society's make-up, even an ant colony. Each is complex with internal and external conditions constantly affecting its state. Each is a *dynamic system*, that is, it is continually changing. Each is *non-linear*, meaning the same set of internal/external conditions do not necessarily produce the same result. Complex systems are constantly in a state of flux as they try to achieve a balance or state of equilibrium tittering between order and chaos. Every

---

complex system has a *tipping point* which when passed can send that system into a chaotic state. The forces inherent within each complex system are constantly trying to adjust its state. Sometimes the result is utter chaos and destruction. Sometimes it is able to rebalance itself and achieve order. **Complexity** is the branch of mathematics which studies and tries to describe complex systems by using mathematical models and computer science.

Chaos theory is a branch of mathematics which studies non-linear systems. Chaos is a state in which nature does not conform to established scientific laws. Before a system becomes chaotic, that is unpredictable, there is a transition point in which errors have been mounting that will eventually render the system chaotic unless changes take place that will swing the system back before the transition point (tipping point) is passed.

---

[1] Since space and resources are finite, population control needs to be seriously and aggressively addressed.

[2] See pages 150-1-151 for facts & figures and details on: *Carbon dioxide levels; The Earth's ocean in climate control; Endless positive feedback loops; the effects on plant and animal life; current effects of global warming.*

[3] See Al Gore's *Earth in Balance* in which he proposes the US role and outlines 8 steps of actions (pp349-351). Not deterred by the US Federal Government's inertia on global warming, New York and eight other Northeastern states agreed to freeze power plant emissions at their current levels and reduce them by 10% by 2020. California, Washington & Oregon are also making efforts together to lower greenhouse gas emissions. These statistics are from *Too Hot to Handle* by Joan O'C. Hamilton. Stanford Magazine: November/Dec. 2005.

graphs—the picture of global warming and postive feedback loops

This graph shows carbon dioxide levels in the atmosphere since 1010. Note the rises since the 1800s, especially over the past 50 years.

A single degree temperature rise may not sound like much, but bear in mind the butterfly effect, in which a minute change can build and set off a global chain of reactions.

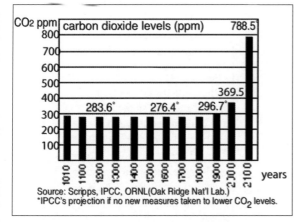

CO2 ppm carbon dioxide levels (ppm) 788.5

283.6°   276.4°   296.7°   369.5

Source: Scripps, IPCC, ORNL(Oak Ridge Nat'l Lab.)
*IPCC's projection if no new measures taken to lower CO₂ levels.

The Earth's oceans play a major role in climate control. They act as a very sophisticated air conditioning system. The seas absorb both heat and carbon dioxide (via their plants) and simultaneously circulate the warm and cold currents and nutrients throughout the globe. Over 70% of the Earth is covered by water, which retains a thousand times

This is primarily due to human industrial activities. The graph also illustrates the sharp increases in carbon dioxide levels in 2000 and projections by the International Panel for Climate Change (IPPC). IPCC (http://www.ipcc.ch/) was established in 1988 by the UN's World Meteorological Organization and the UN Environmental Programme to assess climate change.

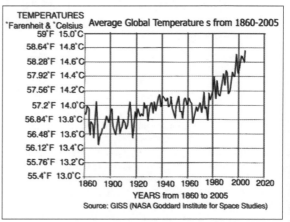

TEMPERATURES
°Farenheit & °Celsius   Average Global Temperatures from 1860-2005

Source: GISS (NASA Goddard Institute for Space Studies)

more heat than the atmosphere. Both marine and land life depend on the health of the ocean.

"Scientists recently detected a weakening of the flow of ocean currents in the Atlantic basin because of an infusion of freshwater from melting sea ice and glaciers. At a certain point, they say, the change in salinity and water density could change the direction of ocean currents, leading to many more severe winters in northern Europe and North America."[1]

The constant rise in carbon dioxide levels seems to create an endless loop of global warming. What's the evidence:

• Photos taken from space show the decrease in the Earth's ice caps. Less white ice means less white area to reflect the Sun's rays into space and more to be absorbed by the land and blue waters which in turn means more carbon dioxide released into the atmosphere. And so the vicious cycle continues. Other signs of temperature increases are:

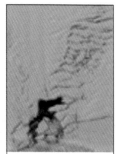

Cracks in the ice show Antarctic icesheet failure. Photo courtesy of NASA.

• The warming rate of the Arctic is 8 times faster in the last 20 years than the previous 100 years.

• 2004 measurements show that the Arctic sea ice was more than 13% below average. It's estimated that the sea ice is melting 20% faster than it did 20 years ago.

• Mount Kilimanjaro's snowcap has decreased by about 80%.

• 120 of the 150 glaciers that existed in Glacier National Park since 1910 have disappeared.

• Three enormous Antarctic ice shelves have collapsed into the sea since 1995. On March 2002 the Larsen B ice-shelf collapsed. It was over 700 feet thick and 1200 acres in area.

As temperature and carbon dioxide levels increase the ocean absorbs more heat and carbon dioxide. This process increases the water temperature making it more difficult for the ocean to function as an efficient natural air conditioner. This means surface temperature will rise further. The potential ongoing cycle is a frightening example of a *positive feedback loop.* The term *positive* does not connote a desirable output, but rather one that increases. A *positive feedback loop* accelerates a process that is occurring, and unless something is done to check it or reverse it, the process will cause a collapse of the system. Positive feedback loops apply to any type of system, such an ecosystem, economic system, political system.

What will happen to the Earth's plant life? Look at the Amazon rainforest. It relies on carbon dioxide for photosynthesis. There's plenty of carbon dioxide since levels are constantly rising, but because of the increase in carbon dioxide the rainforest vegetation won't need to open their stomata cells as often to absorb carbon dioxide. The less they open up their cells the less moisture escapes from the stomata into the rainforest. The Amazon rainforest and river thrive on moisture. As moisture declines, vegetation will dry up, and drying up vegetation releases more carbon dioxide into the atmosphere and again this dangerous cycle perpetuates itself in another *positive feedback loop.*

What about animal life? Animal diversity and species extinction will become inevitable, since food chains will be disrupted. Climate change alters food sources. The ocean's plankton, krill, and the coral reefs will inevitably suffer.

In other words, perpetual cycles increasing surface temperatures lead to more carbon dioxide being released into the atmosphere. Each carbon dioxide molecule caught in the atmosphere is stuck there for a century. This leads to more insulation for Earth which leads to more heat and eventually to a major climate change.

What can we expect from global warming? More global warming. The increases in global warming have already left footprints. A short list includes:

• Sea levels are rising, islands are disappearing, and people have had to relocate. This has happened in Tulun Islands (Piul and Huene islands) near New Guinea. New Zealand has agreed to accept Tuvalu island residents in a timed evacuation plan. As global warming increases more rising sea levels will result which

will create environmental refugees from the coastal regions of China, India, Bangladesh and islands in the Pacific.

• Extreme weather events are already upon us: the 2003 European heat wave killed 27,000 people—between 1999-2004 the western United States experienced its worst droughts in 500 years—in the past 30 years the number of Category 4 and 5 hurricanes has nearly doubled.

• Increases in certain insect-borne diseases such as Lyme disease have occurred because warmer weather is more hospitable to ticks.

• Since temperature record keeping began in the late 1880s, 8 of the 10 hottest years occurred between 1996 and 2005. As of this printing, 2005 is the hottest year ever recorded.

• The permafrost (rock and soil material whose temperature remains $\leq 32°F$ for at least two years) has begun to thaw, especially in Alaska and Siberia. At the end of the last ice age a permafrost area in Siberia spanning a million square kilometers formed 11,000 years ago. It has now begun to melt for the very first time in its history.

• Loss in biodiversity and ecosystem is likely to increase further, especially by 2100.

• Recent satellite data indicate an ongoing shift in arrival times of seasons.

---

[1] From W*hy we need to worry about global warming* by Ross Gelbspan in Perspective section of *San Jose Mercury News*, August 9, 2006.

Archeological digital blueprinting

Making a blueprint requires a variety of measuring tools in addition to the knowledge of converting measurements to scale drawings. When an architect draws plans he/she can do it on paper or computer. But, what about the reverse. What if you want plans for a building that is already built or an ancient temple whose ruins are the only remains. In the past one would have to make thousands of measurements by hand in order to figure out how a structure was made, and come up with plans. Today, thanks to inventor and civil engineer Ben Kacyra, a very special scanning device is available that can make three-dimensional digital blueprints of any structure or archeological ruins with amazing accuracy. Working with UC Berkeley trained engineer Jerry Dimsdale, Kacyra's instrument is bringing the ancient world digitally back to life. MIT had a laser powerful enough for its light to bounce off distant objects while safe enough not to harm eyes. Kacyra was able to license a civilian version of this laser. Los Alamos National Laboratories provided a timing device that could precisely time intervals between the signal and its response. Utilizing these components along with the necessary software, HDD (High Definition Documentation) with HDS(high definition surveying) creates 3-D point cloud models which are a set of points or vertices displayed in a 3-D coordinate system. The points represent the precise location where the laser beam makes contacts with the object. The cloud is composed of millions of points which are reassembled into digital blueprints using specialized computer software along with CAD (computer-assisted-design) drawings and high resolutions photographs. In addition, the cloud point data is accurate to 0.5cm. Kacyra's device along

Mesa Verde, Arizona.

with its software can easily create digital plans and elevations of any structure. It has been used in a number of ways. In restoration of historic buildings it's capable of creating the entire structure so it can be digitally rotated and examined from different views. Ancient structures, such as Mesa Verde, can be integrated with photographs to resurrect the ruins. What's even more exciting is that Kacyra's has set up a non-profit foundation that hosts a website where scientists, students, teachers can access and add information. Log onto http://archive. cyark. org/index.php and be prepared to spend time to explore its many links. John Rick, scientist and expert in pre-Incan Peruvian civilization at Stanford University, points out *"It would take 10 lifetimes to record the data that the Kacyra instrument takes in in 10 minutes."*[1]

---

[1]*Laser Mapping Tool Traces Ancient Sites,* by Tom Abate. San Francisco Chronicle. 7/22/2007.

the computer history museum

Tucked away off highway 101 in Silicon Valley the Computer History Museum attracts computer

The evolution of computer technology has a long and involved history touching many continents and cultures. A quick glance at some of the many tool-like

Tthe Computer History Museum is located at 1401 Shoreline Blvd. In Mountain View, CA.

buffs, students, teachers and phobics from all around the world. If you missed any part of the computer evolution/revolution this is the place to discover/rediscover what it's all about, where it's been and where it's going. Here's your chance to have a look at the amazing developmental journey this incredible tool has taken. See how it seemed to suddenly envelop the world with its peripherals and ever evolving applications.

Today's computers have far exceeded the visions of any one individual. Originally, the term *computer* was used to describe a *person who did tedious calculations.*

"computers" invented illustrates the rich lineage of the modern day computer:

- the abacus was developed and used by many cultures including Chinese, Roman, Japanese, and Greek
- Chinese counting boards and their rods numerals appeared between 2nd century BCE and 2nd century AD
- the quipus were Inca knotted rope computers
- Pascal's calculator made its debut in 1642 while Napier's bones were invented around 1614
- the first slide rule appeared circa 1620

• Joseph Jacquard invented his famous loom in 1801

• Charles Babbage envisioned his differential engine in the early 1800s

• Hollerith's census machine came on the scene with the 1890 census.

The Computer History Museum is an ever evolving display of human ingenuity. Here you can look at an actual model of Herman Hollerith Census Machine along with other artifacts and photos, and learn about this amazing machine's connection[1] to today's IBM corporation. You can see how Jacquard's loom is connected to punch card technology which was used to program early analog computers. There is even a woven photo-like silk portrait of its inventor Joseph Marie Jacquard done on a Jacquard loom. View the mysterious Enigma machine used by the Germans during World War II to encode secret messages, and learn about this and other encoding machines and their connection to computers. View the famous ENIAC computer and other one-of-a-kind computers. Walk down an isle of Cray super computers, and experience the evolution of these super-computers. Explore the many and different calculating machines invented. Delve into the history of computer languages such as AGOL-

60. Come face to face with the Apple-1 computer inscribed by *Woz* (Steve Wozniak). Look at hardware and computer innards of both early analog and digital electronic computers. In addition, the Computer Museum has an elaborate website *(http://www.computerhistory.org)* which allows a visitor to:

• take a virtual tour to many of the permanent exhibits including its visible storage, see highlights of collections, and tour their latest/current exhibits

• join discussions and view global chronicles of the origins, stories and artifacts of the information age

• search their collections of artifacts online

• view the Museum's technical publications

• explore computer restorations being made

• view historical materials about high tech companies

• learn about the sponsored speaker series of well known people in the computer industry and sciences and actually have access to full lectures of past speakers.

The museum and its website definitely merit ongoing visits.

---

[1] In 1911 Herman Hollerith's company merged with Computing Scale Company and International Time Recording Company and formed C-T-R (Computing-Tabulating-Recording) company. In 1914 Thomas J. Watson joined C-T-R and Hollerith retired with $1.2 million sum. In 1924 Watson renamed C-T-R International Business Machines (IBM).

the mysterious hole in m.c. escher's print gallery

M.C. Escher was a

master of transformation both in his incredible tessellations and in his amazing topological works Many of his lithographs give the impression of being printed on

completely understanding the mathematics, nevertheless he intuitively introduced it in *Print Gallery*. Escher pointed out that *"Again I had a most pleasurable contact with some very learned mathematicians, who did their best to explain to me that the print **Print Gallery** has to do with a Riemann's surface. Despite their lectures, my understanding of it is very incomplete."* [1]

Escher's *Print Gallery* with mathematical modification by Henridk Lenstra. The red circle indicates the area which M.C. Escher left blank with only his signature appearing. Image courtesy of escher-droste.math.leidenuniv.nl of Universiteit Leiden, Holland.

In *Print Gallery* we notice a drawing that seems to repeat and twist, yet in its center peering out like an eye is a disconcerting round hole in which Escher placed his signature. Why did Escher leave a blank space smack in the middle of his work? What could logically appear there that would have been consistent with the rest of *Print Gallery?*

rubber sheets. It is among these we find *Print Gallery,* a piece of art which teases and perplexes the mind. Many mathematical concepts are behind the formation of this complex lithograph. Even though Escher himself admits to not

Mathematician Hendrik W. Lenstra of Leinden Universiteit of Holland was intrigued by this blank space, so he set out to mathematically explore and analyze the lithograph. In the process he discovered how and what the blank hole could accommodate that would be complementary with the rest of Escher's *Print Gallery*. He found that although Escher himself did not resort to the mathematics of an elliptic curve over a complex number field, Escher's artistic eye intuitively adapted such mathematics. Lenstra resorted to high level mathematics to explain how Escher produced *Print Gallery* and to discover what could be drawn in its mysterious blank space. Lenstra found that *"...the lithograph can be viewed as drawn on a certain elliptic curve over the field of complex numbers and...that an idealized version of the picture repeats itself in the middle. More precisely, it contains a copy of itself, rotated clockwise by 157.625596082...degrees and scaled down by a factor of 22.5836845286...."* [2] Since the scene is periodically repeated, Lenstra's analysis explains how a smaller copy of the scene ends up appearing in the hole nearly upside down. Then, subsequent periodic rotations of the scene repeat ad infinitum decreasing in size to a point of singularity at the lithograph's center point. The mathematical formula that Lenstra developed to reproduce the grid for Escher's *Print Gallery* and fill in the circular gap is

$$h(w) = w^a = w^{(2\pi i + \log 256)/(2\pi i)}$$

In addition, special software was written by Joost Batenburg, then a mathematics student at Leiden. This software made it possible to reconstruct Esher's studies by working backwards off of Escher's lithograph and grids. Later, artists were also brought into the project to make adjustments and enhancements to the results. Finally, in

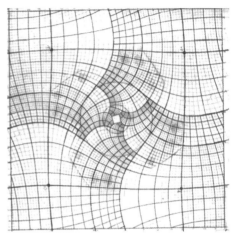

Above is the grid Escher used for *Print Gallery*. Below is the grid that Lenstra mathematically produced for *Print Gallery*.

Images courtesy of escherdroste.math. leidenuniv.nl of Universiteit Leiden, Holland.

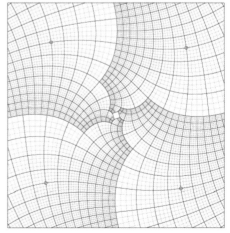

A close-up look at what Lenstra filled in the blank hole of *Print Gallery*.
Image courtesy of escherdroste.math. leidenuniv.nl of Universiteit Leiden, Holland.

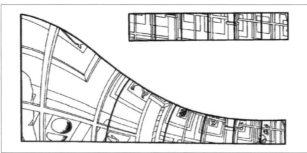

How a straight grid of squares is mapped on a curved grid.

Images courtesy of escherdroste.math. leidenuniv.nl of Universiteit Leiden, Holland.

order to make the grayscales uniform with continuous resolution, adjustments to pixel density were made using various mathematical concepts including the elliptic curve and the exponential function.

To understand, appreciate and fully experience the Lenstra's mathematical analysis of Escher's *Print Gallery* go to the website: *http://escherdroste.math.leidenuni v.nl.*[3] Here you will find an in depth mathematical analysis with diagrams and animations which zoom into the point of singularity of its blank space.

The questions remain: Why didn't Escher fill in the hole? Was the blank hole too complex for Escher to fill in by hand? Only the master knows.

Escher's style, technique, diversity and complexity continue to astound generation after generation. To the mathematician Escher's works abound with mathematical concepts. Among these we find: topology, symmetries, dimensions, impossible figures, perspective, lattice, mazes, tessellation, rotation, translations, transformation. Some of Escher's works would be impossible to create without understanding and following certain mathematical concepts he discovered and applied to this work. For example, Escher's study of tessellations made possible the precision, motion and depth we find in *Metamorphosis* and *Sky and Water*. He not only transforms a plane with his tessellating objects', but even the objects shapes undergo a transformation while tessellating the plane. In some of his works Escher amazingly manipulates and mixes dimensions. For example, in *Reptiles* 2-dimensional lizards eerily come to life in realistic crawling 3-dimensional forms. In *St. Peter's Rome, Tower of Babel* and *High and Low* Escher relies on the concepts of projective geometry and a mixture of traditional vanishing points with his own curved vanishing points to create unusual depth and dimension.

---

[1]From page 88 MC ESCHER His Life and Complete Graphic Works by F.H. Bool, J.R. Kist, J.J. Locher, F. Wierda; Ahrry N. Abrams, Inc. NY, 1981.

[2]From *The Mathematical Structure of Escher's Print Gallery* by Bart de Smit and Hendrik W. Lenstra. *Notices of the AMS(American Mathematical Society*. Volume 50, April 2003.

who's talking math?

For years the general public considered mathematics alienating. Its importance was not fully acknowledged. People just did not want to talk, think, or see mathematics. Many even viewed it with hostility. Today, even though we still hear the infamous phrases *I always hated math* or *I could never do math*, these sentiments are decreasing. Declaring *I can't understand math* is no longer something to brag about. Mathematics is no longer off-limits. References to math and math ideas appear everywhere from gossip columns to ad campaigns to news headlines and in entertainment. No longer mocked or shunned, mathematics is becoming a social force and even a celebrity. We see, hear and read about it:

• in movies[1]...*A Beautiful Mind, Good Will Hunting, Enigma, Breaking the Code, Proof, Conceiving Ada.*

• at the theater... *Arcadia, Proof, Copenhagen, Picasso at Lapin Agile, Fermat's Last Tango*

• on ads plastered on billboards and the sides of buses declaring such slogans as *math makes the difference, do the math, it adds up.*

• in newspapers and magazines

• in fiction...*The Da Vinci Code, Uncle Petros & Goldbach's Conjecture, Case of Lies, Foucault's Pendulum.*

• in wardrobe accessories

• museum exhibits

• comic strips

• And now mathematics even plays a major role on *NUMB3RS*, a prime time TV show whose mantra is *"We all use math everyday: to predict weather, to tell time, to handle money. Math is more than formulas and equations. It's logic. It's rationality. It's using your mind to solve the biggest mysteries we know."* The program doesn't hesitate to bring in and use high powered cutting edge mathematical concepts to help the FBI solve cases. Math ideas such as *game theory, probability, origami, Bayesian filters, combinatorics, supersymmetry, 3-D scatter plot, decision theory, pi, quantum entanglement, data mining, Fourier analysis, emergence theory, triangulation* and many more are discussed, explained and used to solve crimes and/or explain human behavior. All these ideas are presented in a natural and positive way connecting them to real life experiences.

What's the catalyst behind this mathematical trend? Perhaps it is the wide spread use of computers and computer related technology. Or, is it because people now realize that other mathematical concepts besides numbers and computation are pervasive in their daily lives?

Mathematics and logic go hand in hand, and mathematical applications have come into the limelight. Among the many ideas we find are:

• *in medicine.* People are finding that they need to be informed now more than ever about their health numbers (*blood pressure, cholesterol ratios, diabetes numbers, blood coagulation -INR numbers*), about new imaging technologies and how they work and don't work in order to make informed health decisions.

• *in new modes of communications and electronics* such as cell phones, emails, iPods, audio and video streaming, chat rooms, computer telephoning, GPS units, PDAs all of which require logic to operate. Today's electronic gadget world requires a general understanding of math and logic.

• *behind almost every facet of our modern lives* from tracking financial portfolios to economics to politics to sociology.

To function in our high tech world we are forced to deal with numbers to protect our privacy be they PIN numbers, ATM numbers or passwords. We rely on numbers as our protective armor. Even our automated garage doors require a numerical code. Numbers and math ideas are used to quantify and describe just about everything including the size of earthquakes, the category of hurricanes, the

prime rate, a new born baby's social security number. The word *help* has become synonymous with the number 911. More and more we find ourselves less able to function without numerical data. We need precise

numerical data and logic to make informed judgements. We can no longer get by without knowing some math. Insurance companies, pharmaceutical drug trials, population growth, profiling, advertising —the list goes on and on — all rely on the mathematics of statistics, probability, analysis and especially the ideas of complexity theory. We've come to realize that everything is interconnected. *All areas of science are interconnected to one and other, and are stitched together by mathematics.*

Each new discovery, each new connection to our daily lives brings us closer to appreciating mathematics. Mathematics, science, and logic are critical to technology. Without getting into them, even though peripherally, one will be left *"out of it"*. For the first time in history, the *"in"* crowd is the math crowd.

---

1 There are a number of websites listing movies with mathematics in them, for example http://std.com/~reinhold/mathmovies.html.

mathematics uncovers the invisibility cloak

Whether it's Harry Potter's invisibility cloak or the Romulans' cloaking system on *Star Trek*, cloaking has been a concept reserved for fiction and science fiction. Not any more. Mathematician Sébastien Guenneau of the University of Liv-

Light can actually be made to bend around objects made with metamterial such as a jet and render it invisible.

erpool in the UK developed a computer model which proves that light can actually bend around an object when cloaked with *metamaterial*. Guenneau explains *"Using this new computer model we can prove that light can bend around an object under a cloak and is not diffracted by the object. This happens because the metamaterial that makes up the cloak stretches the metrics of space, in a similar way to what heavy planets and stars do for the metrics of space-time in*

Einstein's general relativity theory."[1] If we do not see the light bounced off an object, we cannot see the object. Just like we cannot see objects that exist in a dark room. An object draped or whose outer layer is made from metamaterial would bend the light around it to its opposite side, so in essence we end up seeing what's behind the object. It's as if its shape were hallow and our line of sight passed through it. Consequently, it would not cast a shadow.

What is metamaterial? The term *metamaterial* first appeared in 2001 in a paper written by Rodger Walser, professor at the University of Texas at Austin and director of its Center for Electromagnetic Materials and Devices (CEMD). He described it as an artificial composite material that performs beyond the limitations of conventional composites. Various metamaterial possess diverse and unusual properties. The designs are three-dimensional with periodic cellular structure. DARPA (Defense Advanced Research Project Agency) Metamaterial Program describes metamaterial as a new class of composites whose orderly design exhibits exceptional properties not seen in nature. The properties of metamaterials come from their architectural design rather than from their composites.

Until metamaterials, camouflaging and optical illusions were the only means of making objects appear invisible. These included such techniques as —the radar absorbing dark paint of stealth aircraft—optical camouflage for minimizing electromagnetic emissions by cooling the exterior surface of an object—jamming devices that inhibit sensors. Today DARPA is seriously exploring the use of light bending properties of metamaterial to prevent an object from diffracting light.

Metamaterial does not just work to make things invisible. Scientists are also studying *superlenses* which will be able to focus light to a spot much shorter than its wavelength. Visible light's normal visible limit is about 600 nanometers. Roberto Merlin of the University of Michigan in Ann Abror is working on designing metamaterial for superlenses that would both be easier to make and would focus light to a much shorter spot at about 500 times smaller. His design consists of a plate with concentric circles in which the rings alternate between material that blocks light(e.g. metal) and material such as plastic that lets light pass through it. The amazing thing

about superlenses is that they can bend light in the opposite direction of the lens. Such lenses can be used to produce ultra sharp detailed medical scans.

The same science of metamaterial also applies to sound waves. Guenneau and his team designed metamaterial to use with sound waves, and produced a mathematical model showing how it would function. Such metamaterial can bend sound waves backwards, and would be invaluable for soundproofing. The production of sound dampening devices would mean major improvements in acoustics. These metamaterials could also be used to make detailed seismic maps and be helpful in the design of earthquake resistant buildings.

With regard to invisibility cloaking Guenneau explains *"humans beings and animals are more difficult [to cloak] as their movement is very flexible so the cloak...would easily be seen when the person or animal made any sudden movement."* But that does not mean it's impossible. In fact, researchers predict progress within the next decade.

---

[1] *Cloaking Device? Invisibility Technology one Step Closer.* Science Daily, May 4, 2007.

mathematics, physics & art all in one
the art of michael burke

Who would expect to find the language of quantum physics in art? Yet, who expected to see the theory

Michael Buck seven sculptures composing *The Quantum Stream* were inspired by quantum mechanics.

of relativity in Salvador Dali's *The Persistence of Memory* or multidimensions in Pablo Picasso's *Man with Violin*. Michael Burke—artist and former astronomer and city planner with a background in physics, architecture and mathematics—creates dazzling aluminum sculptures which speak of mathematics and science. When most people think of art they usually do not associate it with mathematics or science. Yet artist Michael Burke considers them *"one and the same"[1]*. In his *Quantum Stream* one can conjure up quantum particles of light streaming by at the speed of

light. If these were on the Internet they would be carrying packets of information, but instead these "quantum particles" are large sculpted aluminum shapes stretching across a landscape each carrying its own mathematical/scientific message. Each sculpture has been ground in such a way to intensely collect and scatter light. *Quantum Stream* consists of seven aligned sculptures modeling how light travels in a straight line, yet all these "light particles" have their own unique shapes and formulas. Burke describes these sculptures as reliquaries, and explains that *"each has a scientific relic (formula) within"*. For example, the close-up of *Photosynthesis* shows its own inscribed formula and symbols. Among some of the other formulas in *Quantum Stream* are the expansion rate of the universe and Fibonacci numbers. Burke emphasizes that neither the sculptures' shapes nor their formulas are to be taken literally, saying *"each (sculpture) represents a quantum of light and each possesses a different scientific concept"*. On some of his pieces the formulas are written upside down, sideways or even reversed since understanding the formulas is not as important as sensing the connection between science and art.

His *Neutrino Collector*, was originally installed in an ancient Etruscan tomb in Italy, and in essence represents the process of collecting neutrinos which takes place deep in the earth using heavy water. The mathematical equation inscribed on the sculpture is that for a neutrino[2] reaction with a whimsical twist—Burke purposely left a blank space where the term for the neutrino would appear bidding an actual neutrino to sit and balance the equation.

Etruscan script and quantum mechanics formulas are inscribed on small aluminum sculptures on the floor of this Etruscan tomb which surrounds Michael Burke's sculpture, the *Neutrino Collector*.

A close-up view of the Photosynthesis quantum sculpture of *The Quantum Stream*. The scientific and mathematical symbols deal with respiration and transpiration of the O_2 and CO_2 interchange during photosynthesis.

When laid out flat the shapes on the pages line up in a certain way so that some parts of the shapes can be seen through the various layers of the book. Again one finds scientific writing and formulas on the book's pages. For example, for the *Light Waves* book, Burke explains that this *" book is about light...within the book there are formulas that have to do with the*

A series of Burke's sculptures are aluminum books ranging in size from eight inches to six feet. They can be "read" like a book whose pages are bound with aluminum hinges. Pages consist of cut out forms. When stood upright these books are transformed into sculptures.

Above, *Light Waves,* one of Burke's aluminum book sculptures. Photograph by Jonathan Smith.
To the right is a six foot version of *Light Waves*.

A close up of the first sculpture in the *Quantum Stream* with the formula for generating the Fibonacci sequence.

remember the book. Burke explains that although the pages of a book are usually read sequentially, one's memories of a book are often fragmented, and most often not recalled sequentially.

It is not Burke's intention that the shapes of his sculptures convey scientific ideas, but their etched formulas convey in a subtle way that scientific ideas are inherent in art. When the observer looks closely to discover the language of physics incised on his art forms, the formulas are not meant to stand out, but rather they are intentionally burnished and obscured explaining that by their very nature they are difficult to comprehend.

transmission of light. The way it (the aluminum) is ground scatters light all over...the light almost interferes with you seeing the book...it forces you to recognize that the only way you're seeing this book is through light". For a book installation, Burke prepares ink rubbings of the pages' surfaces. He combines various parts of these rubbings in a variety of ways, and frames various combinations on aluminum type mattings that have also been ground and etched with formulas. Burke further explains that the *"aluminum scatters light around it so ... that the paper (ink rubbing) on it seems to float (above) the aluminum (sheet)".* Another method Burke uses to enhance a book's exhibition can be seen in the photograph of the exhibit at Rockefeller University. Here the aluminum book is surrounded by rubbings of its pages highlighting Feigenbaum's universal constant. The seven pictures comprising this exhibit represent some of the many ways one can

Rockefeller University installation (2004) was part of the University's celebration of the work of Mitchell Feigenbaum's contributions to chaos theory. Burke's four page aluminum book on the right is etched with Feigenbaum's universal constant[3] $\delta = 4.6692016091029906...$ flowing off its pages.

In the foreground is one of Burke's aluminum obelisks which has numbers, pieces of constants cut into a hollow aluminum beam.

Michael Burke's work combines his imagination with his scientific and architectural knowledge into sculptures that bring to our attention the innate connection and relation between art and the sciences.

A close-up of *Obelisk I.*

Having grown up in a home surrounded by art and literature, Burke was exposed to many disciplines. In college he studied mathematics, architecture, astronomy and art. A vivid memory he still shares from college days was realizing the art and beauty in the shapes and forms of mathematical writing and formulas covering the blackboard in his advanced calculus class. After graduating from Harvard in architectural science, he worked for the Smithsonian Astrophysical Observatory which sent him off to astronomical observatories around the world. Yet art always remained a part of his personal focus. In 1967 he decided to change careers, and entered Columbia's Urban Studies graduate program where his interest in architecture resurfaced. After graduating from Columbia he taught classes there in urban planning and environmental studies in addition to working in city planning. Although Burke always was involved in art, it was not until 1975 that he decided to pursue art fulltime. His artwork now primarily focuses on sculpting in aluminum, and is in both public and private collections.

[1] Unless otherwise indicated quotes are from a November 2007 interview with Michael Burke.

[2] Neutrinos are elementary particles, that is particles not known to be composed of smaller particles. Neutrinos lack an electric charge, travel at nearly the speed of light and can pass through matter virtually undisturbed, and consequently are very difficult to detect. There are several types of neutrinos and antineutrinos—electron and muon to name a few. Most neutrinos that pass through the Earth come from the Sun with trillions passing through the human body every second.

[3] The Feigenbaum constant is a universal constant discovered by Mitchell Feigenbaum. This number is common among a class of chaotic systems, similar to how π is a common number for the class of geometric objects known as circles. For complex systems which bifurcate and have periodic doubling the Feigenbaum number indicates when to expect the next bifurcation, and following ones—in other words, when to expect a complex system to become chaotic.

The Archimedes palimpsest

W hat was so special about a very old, very shopworn, moldy scorched book, and who would pay millions of dollars for it? Its story reads like a mystery fulls of intrigue, greed and fraud—almost as fascinating as the

tician and historian Johan Ludvig Heiberg came across a parchment book whose obscure background writing hid some of the work of Archimedes. Sometime during the 10th century Archimedes' work had been transcribed. We also know that in the 13th century the parch-

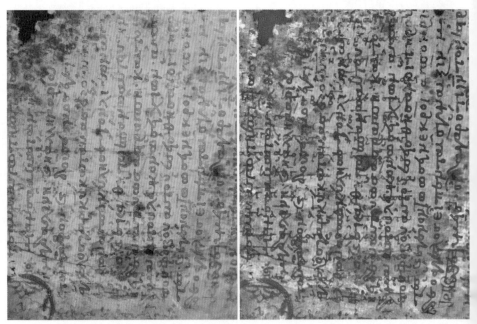

Two photos of the same portion of the Archimedes Palimpsest. On the left is an unenhanced photo. The processed image on the right clearly shows the vertical and horizontal writing. This portion contains parts of Archimedes' treatises on the *Measurement of the Circle* and *the Stomachion*. Image taken by the Rochester Institute of Technology and John Hopkins University. Copyright resides with the Owner of the Archimedes Palimpsest.

contents of the palimpsest. Hidden within its pages were the writings and thoughts of Archimedes. It was the Archimedes palimpsest.

In the early 1900s at the library of the Church of the Holy Sepulcher in Constantinople, Danish mathema-

ment of the manuscript was recycled by Byzantine monks, who scrubbed and scraped the pages of parchment. Such recycled erased parchments are referred to as *palimpsests*. In this case the recycling process involved unbinding the pages of the manuscript and removing as much of the copy as possible. After the folios had as much writing removed as possible, they were cut in half, and the pages turned 90 degrees and stacked . These "blank" pages were then written over in lines vertical to the original horizontal lines of Archimedes' text. Why would such a valuable manuscript be recycled? Most likely because the monks and scribes did not understand the scientific and mathematical information and its significance. To them the value of the parchment exceeded the value of the "meaningless" content. Thus, the monks reused the parchment to write a religious text and then rebound the book.

Archimedes' work within the parchment book was barely visible to Heiberg. He photographed pages and began transcribing it using only a magnifying glass. Unfortunately the book was stolen before Heiberg and others could thoroughly examine and complete the transcription.

Around the 1930s a French collector purchased the text in Constantinople. It was later ascertained that a forger had apparently painted gold leaf religious images on four of its pages in hopes of enhancing its value. The manuscript remained with the collector's family until 1998 when the family sought to sell it through an auction at Christie's auction house in New York. At this time a lawsuit was filed by the Greek Orthodox Diocese of Jerusalem contesting ownership. The judge ruled that the French family possessed title and the right to sell the palimpsest. Archimedes' work was auctioned off for over $2,000,000 and was purchased by an anonymous American billionaire who made it available to the Walters Art Museum in Baltimore, Maryland where scholars and scientists could study it. With the use of modern technologies, ultraviolet lighting and computer programming Archimedes' text, heretofore obscured by the religious text, was dramatically revealed. The palimpsest was found to include seven treatises by Archimedes: •(1) *Equilibrium of Planes* •(2) *Spiral Lines* •(3) *The Measurement of the Circle* •(4) *Sphere and Cylinder* •(5) *On Floating Bodies* •(6) *The Method of Mechanical Theorems* •(7) *the Stomachion*. Up until now, there was not an extant manuscript or copy of *On Floating Bodies* written in Archimedes' Greek, and no versions existed in any language of *The Method of Mechanical Theorems* and the *Stomachion*. In addition to the Stomachion puzzle and mathematical text, the palimpsest includes over 50 mathematical diagrams including the famous *Archimedean spiral*. This palimpsest is an invaluable source of the ideas and new nuances of Archimedes' scientific thinking.

For the *Archimedean spiral* the distance between coils remains constant. Its equation written in polar coordinates (r,θ) is $r=a+b\theta$ where r and θ are real numbers Changing a's value turns the spiral, and b's changes affects the constant distance between the coils.

the stomachion puzzle

A portion of the Archimedes palimpsest includes the geometric puzzle called the *Stomachion,* one of the oldest known puzzles. The puzzle's problem is to arrange its 14 pieces to form a square and other geometric figures as well as figures of animals and other objects. What was Archimedes (287-212 BC) doing

Speculations have abounded ever since a manuscript containing this puzzle was discovered in the contents of the palimpsest. Among the mathematicians and scientists who have scrutinized the Stomachion puzzle are: mathematics historian Reviel Netz (Stanford University), classics professor Nigel Wilson (Oxford University), mathematicians

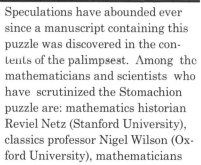

Persi Diaconis (Stanford University), Susan Holmes (Stanford University), Ron Graham and Fan Chung (University of California, San Diego), and computer scientist William Cutler. Since Archimedes felt the puzzle was important enough to write down, he must have been analyzing it. Here's what mathematicians uncovered.

One of the 536 distinct ways the Stomachion's 14 pieces can be arranged into a square.

• If the puzzle is placed on a grid, each of 14 pieces' vertices always land on points of the grid regardless of the arrangement.

• When the area of each piece was calculated it always was an integral fraction of the square's area. For example, diagram 1 shows the pieces that compose a 12x12 square. The number appearing inside each piece is the

with the Stomachion mathematical puzzle over 2200 years ago? Was it merely a form of recreation for him? Was he using it to delve into mathematical concepts? Was he trying to formulate a mathematical proof?

piece's area. The area of the entire square is 144 square units. The pieces' area are each integers — 3, 3, 6, 6, 6, 6, 9, 12, 12, 12, 12, 12, 21, 24, making each fractional amount of 144, namely 1/48, 1/24. 1/16, 1/12, and 1/6.

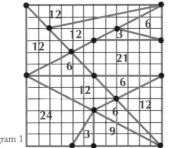

diagram 1

• Pick's Theorem can be used to find the area of the puzzle's pieces. *What's Pick's Theorem?* Discovered by Georg Pick in the late 1800s, his theorem gives a neat ingenious formula for determining the area of a simple lattice polygon (a polygon whose sides do not intersect other than at the vertices and whose vertices lie on points of the grid). If I=the number of interior lattice points, B=the number of grid points on the polygon's boundary, then the polygon's area is given by the formula: $area = I + (B/2) - 1$. Here is how it applies to the shaded piece of the puzzle shown in diagram 2.

area of shaded polygonal piece=
$$7 + (12/2) - 1 = 12$$

• Initially it was found that the puzzle's pieces could be arranged in no less than 268 distinct ways. In 2003 Cutler, using computer technology, showed there were 536 dis-

tinct ways to tile the 14 pieces into a square, and produced illustrations of all 536 tilings. This number excludes any tilings which are rotations or reflections of others. With rotations and reflections there are 17,152 different ways of arranging the Stomachion pieces into a square. (See http://www.maa.org/editorial/mathgames/ mathgames_11_17_03.html to see all 536 of Cuter's solutions.)

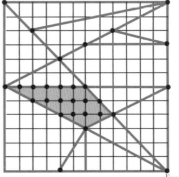

diagram 2

• Note in the diagram below there are three sets of differently shaded pieces. Each piece of a pair always occurs adjacent to the other in all 536 arrangements.

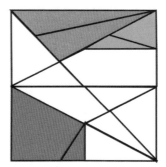

• Whenever the arrangement of the tiles had a symmetrical object like a square, a rectangle, or isosceles triangle, it could be flipped to produce a new tiling of the square, as shown below.

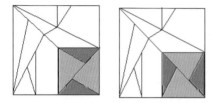

• Most of the different tilings are connected to the traveling salesman problem[1]. In other words, one can transform one tiling into another in less than a certain number of minimum steps.

Mathematicians have speculated how Archimedes was using the puzzle. Some feel Archimedes may have been looking for another proof of the Pythagorean theorem. Some propose he may have been using it to calculate areas of complex polygonal shapes by forming the pieces into a square with a polygonal hole. By trying to find the different ways to tile the square, others speculate he may have been one of the first to delve into the relatively new branch of mathematics now called combinatorics. Diaconis points out that *"Combinatorics is the Johnny-come-lately of mathematics, and mathematicians don't treat it with the re-* *spect they give geometry or number theory. This shows we have lineage too."* [2]

What we do know for sure is that this puzzle is a treasure trove of mathematical ideas with other concepts still waiting to be uncovered. How many of these ideas Archimedes had considered remains pure speculation. Hopefully the Archimedes palimpsest may provide some inkling as to what type of mathematics Archimedes' fertile mind was exploring.

It took genius from many disciplines—history, language, mathematics, engineering, computer science, etc.— to piece together the fragmented information from the Archimedes palimpsest. A major break through in making the Greek text visible came from an idea by staff scientist Uwe Bergmann at Stanford Synchrontron Radiation Laboratory. Bergmann and his team at SSRL were able to significantly enhance pages of Greek text that other methods failed to do. Learning that Byzantine scribes made ink by combining oak galls and ferrous sulphate, Bergmann realized that perhaps the Greek writing would stand out if the synchrotron could detect the iron particles in the ink. His team successfully improvised means to use the synchrotron to bring out the hidden writings.

---

[1]Problems involving the shortest distance, time or least cost of having to travel between various destinations.

[2]*Glimpses of Genius* by Erica Klarreich. *Science News*, May 15, 2004.

mathematical origami

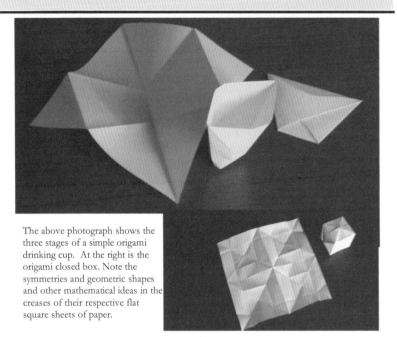

The above photograph shows the three stages of a simple origami drinking cup. At the right is the origami closed box. Note the symmetries and geometric shapes and other mathematical ideas in the creases of their respective flat square sheets of paper.

The term origami conjures up images of amazing sculptures created from a single square sheet of paper. Dating to 583 AD when Buddhist monks brought paper into Japan from China via Korea, origami evolved into an art form with links to various cultures and countries. Over the centuries, thousands of objects have been created restricted only by the imagination.

Origami impacts our lives in subtle ways, appearing in folded napkins for special table settings, in household decor and jewelry, in architectural forms, and even in the way automobile airbags[1] are stored.

Many properties and figures from Euclidean geometry manifest themselves in the creases and the valleys of the unfolded square of the origami model. Triangles, polygons, congruent figures, ratio and proportions, symmetries and similarities appear in the creases. Some models reveal iterations, tessellations and even fractal-like creases.

Mathematics and origami connections have been evolving for decades. As the study of mathematics and paperfolding is being explored so is origami itself.[2] Today mathematicians are using origami in new and surprising ways. The crease formations on the flattened square are studied, analyzed and applied to such areas as graph theory, combinatorics, optimization problems, fractals, topology, and supercomputing.

Mathematicians are developing an entire system of origami mathematics with its own definitions, concepts, theorems, models, and applications. Here we find definitions for flat folds, vertices and mountain and valley creases along with their very own symbols. This symbol — . . — . . — . . — . . — is for the mountain crease and — — — — — for valley creases.

> **flat folds** or **flat origami models:** origami models that can be flattened without adding any additional creases. These collapse effortlessly from 3-D forms to 2-D forms.
>
> **vertices:** the point at which two or more folds meet
>
> **mountain crease:** a crease that is folded upward

Some fascinating theorems which analyze, predict and generalize what takes place in the folds of origami objects are:

• *Maekawa's Theorem*[3] points out that the difference between the number of mountain creases and the number of valley creases in a flat vertex fold is always two.

• *Kawasaki's Theorem* shows that the sum of every other angle about a flat origami vertex in a flat fold is always 180°.

• *The two color origami map theorem*, reminiscent of the four color map problem, proves that various regions formed by creased patterns of a flat folded origami model can be distinguished using only two colors[4].

• A jazzy theorem by computer scientist Erik Demaine points out that *an origami square sheet of paper which is black on one side and white on the other can be folded into a checkerboard pattern.*

A new area being explored in origami is known as *rigid origami*. Rigid origami studies and analyzes origami models which can be made from rigid materials such as cardboard or sheet metal. These models must be folded without bending or twisting the material; consequently, they fold along hinges rather than creases. Rigid origami delves into designing flat formed objects that can go effortlessly between their 2nd and 3rd dimensional forms, creating models that can be easily flattened or pulled into firm 3-D forms. Imagine the portability of such models! They could be transported easily, and no assembly would be required on the other end. Can you envision origami dwellings or modular origami forms introducing a entirely new type of architecture?

Where is this origami mathematics leading?

• Computer programs such as — *TreeMaker* by Robert Lang helps in the design of complex origami models —*Tess* by biochemist Alex Eaterman can theoretically design flat-folding origami models with infinitely many creases and create origami patterns which tessellate (See Bateman's website: http://www.sanger.ac.uk/Users/agb/Origami/Tessellation/ where you can download Tess.).

• The *Muira Ori* folding pattern is used in retractable solar panels for artificial satellites. In addition, Koryo Miura (professor emeritus at University of Tokyo) developed a special map folding technique which allows a map to be easily opened and closed as a whole or in sections.

• Nature has been way ahead of science and origami in the art of folding. Much has been learned by observing how birds fold their wings and petals and leaves unfurl.

• The creation of a very large origami telescope (100 meter in diameter as opposed to the 2.5 meter of the Hubble telescope) is currently under way at Lawrence Livermore National Laboratory.[5] The main problem was how to fit such a huge telescope in the space available in the transport rocket. An origami design is the answer. It will function somewhat like an umbrella, which will not be opened until it is put in orbit and instructed to open.

• Rigid origami containers. One such is the Fuse pot designed and patented by Tomoko Fuse of Nagano, Japan. Fuse created her sturdy

traditional origami rules: In traditional origami, one is restricted to using only a square sheet of paper, no scissors, glue or tape. Objects must be created by folding a square sheet of paper.

mathematics found in origami: Origami transforms a square sheet of paper into thousands of different 3-dimensional forms—from boxes to birds to rings— which are reminiscent of some of the transformations of topology.

In addition, the crease patterns formed on a folded square can reveal patterns of symmetry, fractals and infinity.

graph theory: Here a problem or object can be reduced to a network of nodes and bridges. For example, a 3-dimensional cube can be reduced to this 2-D graph. An origami figure can also have a planar graph associated with it. The complexity of the origami figure increases the complexity of its graph.

combinatorics: deals with studying the various combinations of things, such as votes in election, travel routes

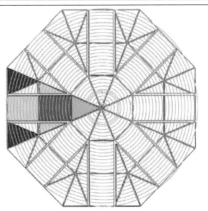

The above is the design behind the 5 meter prototype origami telescope. It is composed of 8 "petal" shaped panels with each consisting of three isosceles triangles, four right triangles, and two rectangles. One of petal's components is shown in various colors. The concentric circles suggest some of the 19,105 circular etched grooves that focus the light. Diagram courtesy of Lawrence Livermore National Laboratory.

origami pot from a single sheet of paper that snaps into a pot or flattens for storage without the use of any adhesives. She envisions that if these pots were made out of durable coated paper, they would be environmentally friendly and economically feasible alternatives to plastic disposable containers.

• An origami bucky ball, two feet tall, was made from origami modulars by mathematician Thomas Hull at Massachusetts's Merrimack College. Hull has done extensive work on origami math uses in graph theory and has proven numerous theorems in this field. (See website: http://www.merrimack.edu/~thull/projectorigami/toc.html.)

The amusing and very old pastime of origami continues to evolve as an art whose applications reach into science and mathematics.

for a journey, the various ways three cards can be drawn from a deck, or even the various ways creases can appear on a folded piece of paper.

optimization problems: These problems seek solutions which most efficiently and effectively use available resources.

supercomputing: Here analogies of complex origami paper folding are explored and applied in factoring prime numbers.

Recall the three impossible construction problems of antiquity—trisecting an angle, duplicating a cube, squaring a circle—using only a straight edge and compass? Origami offers a new approach to bisecting angles, making regular polygons, or making perpendicular bisectors—all without a straightedge or compass. Using its folds and creases, origami even reveals a way to trisect an angle and duplicate a cube.

---

[1] Methods used to design complex 3-dimensional origami objects have been applied to modeling folding patterns for airbags. Rainer Hoffmann of EASi Engineering in Alzenau,Germany, and origami designer Tomoko Fuse of Nagano, Japan, have worked in this area using computer simulations and engineering for the automotive industry.

[2] For example, in 1934 Princeton graduate student Arthur Stone and his colleagues Bryant Tuckerman, Richard Feynman and John Tukey introduced us to the mathematics behind paperfolding hexa-hexa-flexagons. They studied and developed the mathematics behind how to fold a long rectangular strip of paper into an hexagonal object with six distinct flat faces which are revealed as the object undergoes a series of flexings.

[3] Although this theorem was also discovered by Jacques Justin in 1986, Maekawa's Theorem was named after Japanese physicist Jun Maekawa, who created many complex origami models. This theorem seems analogous to Euler's formula relating the vertices, edges and faces of a polyhedron, $F+V-E=2$.

[4] The Four Color Map Theorem states that only four colors are needed to distinguish the regions of a map. It was proven by K. Appel and W. Haken in 1976.

[5]Check https://www.llnl.gov/str/March03/Hyde.html to view prototype and photos.

With today's more than 2 billion websites[1] constantly growing and storing raw data, the task of how to organize, access, transmit and utilize information is an ever evolving ed. Some may think such ideas always existed, but alphabetizing dates back to the Library of Alexandria (circa 280 BC) and the Dewey Decimal System was invented by Melvil Dewey in 1876.

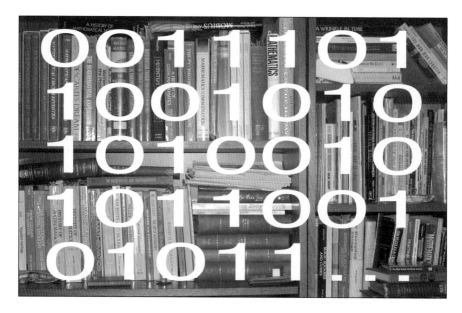

process which relies heavily on mathematics. We are accumulating and storing data at a phenomenal rate! Consider the amount of raw data available through the Internet— every terabyte, megabyte, kilobyte, byte and bit piece of data—and add to this the data that is continually being created and digitized. How does one access what one is looking for in such an enormous changing stockpile of raw data? Today we take alphabetizing and chronological listing for grant-

In the past, data mining was a means by which scholars and researchers used the Internet to rapidly communicate and research their work, ideas, and findings. Today computer scientists and mathematicians are developing new techniques and refining old ones in efforts to meet demands of individuals, businesses, agencies and governments. *Data mining goes beyond just accessing information.* It also deals with using data to predict and explore outcomes. For

A *mathematical lattice* is a set of objects with one or more relations or operators. The relations and operators partially order the set's elements. A mathematical system is created using the set, its relations and its operators so that axioms, definitions, theorems evolve. The mathematical lattice derives its name because its graph may resemble the shape of a lattice. Examples may resemble these:

One can think of *lattices* as configurations of objects on a plane or higher dimensions. In mathematics there are many types of lattices. The point lattice is a regular array of points on a plane, such as a grid of points on a Cartesian coordinate system.

The complexity of the lattice's shape depends on how involved or connected the various points are to one another.

wise go undetected. Fields such as transportation, finance, communication, manufacturing, retail, and health care use data mining to improve customer service, sales and promotions. For example, sales data is used by companies such as Amazon, AT&T and American Express to analyze consumer behavior in order to create promotional strategies. However, there can be data pitfalls especially if the data generated from data mining is too sparse to draw conclusions or make accurate predictions. Regardless of how many mathematical methods are used to tweak the data, the analyses of the data can end up being unreliable and its conclusions may be flawed. Statistics and probability require large samples to be able to make sound predictions. For example, just consider tossing a coin. Probability tells us the odds for getting heads is 50:50, and that the more you toss the coin the more likely it may reach these odds. In ten tosses it may happen that all come up heads, while with a thousand tosses the likelihood of approaching the odds 50:50 is much improved. In a branch of mathematics called *lattice theory*, the idea of *formal concept analysis* is used to conceptualize or create a diagram of the data. Data miners can use *formal concepts analysis* to profile criminals, terrorists, or customers. Here a lattice (a special type of graph) is used to help interpret data. The points or nodes demonstrate connections where interpret-

example, it may be used to reach conclusions, suggest campaign strategies for businesses and candidates, or enable financial institutions to detect fraud. It not only uncovers knowledge, but can analyze it to predict outcomes, create advertising, or pinpoint and entice customers.

Data mining and data analyses go hand in hand. Mining data utilizes mathematical tools that shift, sniff, and sort data. These tools look for patterns and apply statistical analyses to the results. Often information is uncovered that might othering and understanding of results is crucial. An example of misinterpretation of data is sited by Jonathan David Farley in New York Times, writing *"Jafar Adibi, an information scientist at the University of Southern California, analyzed e-mail traffic among Enron employees before the company collapsed. He found that if you naively analyzed the resulting graph, you could conclude that one of the "central" players was Ken Lay's ... secretary."*[2]

Businesses use data mining to target certain customers by data mining their sales records. They analyze customers' buying patterns and preferences, and when a customer goes on line they can suggest other products that might be of interest. They can also use the mined data to design promotions such as discounts when two similar items are bought. This type of relationship grouping is referred to as *association rules* in transaction-based data.[3]

Like other sciences, data mining has evolved its own vocabulary. Among its terminology we find: *bagging, dredging, fishing, metadata, predictive data mining, drill down analysis,* and *boosting.* The data being mined today includes words, numbers, images, photos, fingerprints, diagrams, sounds, etc.. The data miner is actually not much different from a prospector who looks for signs and knows what geological conditions can lead to certain types of ores. The data miner studies patterns in raw data, writes programs to shift and sort by using various branches of mathematics[4], collects samples, creates computer models, studies clusterings, then looks for new patterns and new tools to explore, make predictions, or reach conclusions. Although data mining is a relatively new field it is already big business. It still caters to researchers, scholars, individuals, but it often strikes gold when helping businesses generate new business and sales.

---

[1] The actual number is very difficult to derive with all the various countries and languages. Not all may be registered locally. A study conducted by Inktomi and NEC Research shows that there are over two billion websites.

[2] *The N.S.A's Math Problem* by Jonathan David Farley, New York Times, May 16, 2006. Farley is a science fellow at the Center for International Security and Cooperation at Stanford University.

[3] The famous Walmart *beer & diaper* data mining association in the 1990s supposedly uncovered a link between male customers who purchased beer and diapers on Friday nights. As a result Walmart displayed premium beers near the diapers. Such a move significantly increased their sales of premium beers. This type of data mining is often referred to as *market basket analysis.*

[4] Among the fields of mathematics being utilized are: artificial intelligence, computer sciences, algorithms, probability, statistics, lattice theory, combinatorics, parallel and distributed computing, bioinformatics, discrete mathematics, genetic algorithms, optimization, risk analysis. applied mathematics, and topology.

mathematics fine tunes weather prediction

Using mathematics and physics, our daily weather forecast may eventually become a daily weather fact rather than a prediction. One of the most difficult and accurate predictions to make is pinpointing the time and place of regional rain. Will it rain in my town at 12 midnight or 12 noon? A recent discovery by physicists at Weizmann Institute of Science in Israel may have uncovered a new major player in rain forecasting. How does rain develop? When the

es on minuscule-sized dust particles (about a micrometer) floating in the atmosphere. These eventually form into millimeter-sized droplets. These droplets accumulate more moisture and grow into raindrops large enough for gravity to tug them back down to the Earth. It sounds like a clear logical explanation, doesn't it? Wrong!

Scientists and meteorologists know many other factors come into play. Here are but a few parameters[1] their equations and formulas use:

n(a) cm—*Distribution of droplets over sizes in a unit volume*

a—*Droplet radius*

t—*Time*

r—*Spatial coordinate*

q —*Rate of condensational growth*

K —*Collision kernel*

u_g —*Terminal fall velocity*

g —*Acceleration of gravity*

n —*Air viscosity*

ε —*Energy dissipation rate per unit mass*

u(r, t) —*Air velocity*

earth is heated, formation of water vapor from both land and sea rises as the vapor is warmed. As vapor rises, it begins to cool and condens-

As a result, describing and predicting an accurate rainfall event is not easy. Although meteorological information and techniques have improved dramatically over the decades, the complexity of weather systems, as illustrated by the famous butterfly effect and the Lorenz attractor, produces surprises. Meteorologists know that wind velocity can increase the coalescence of moisture into raindrops and speed up their formation especially when they begin to fall. Even so, predictions of rainfall prove difficult to pinpoint. Physicist Gregory Falkovich of the Weizmann Center's Department of Physics and Complex Systems in Israel points out *"When you estimate the typical time you need to grow from micrometer to millimeter-sized droplets, it would take maybe ten or fifteen hours. And empirically, people noticed that often rain starts long before this—say in half an hour."*[2] That is quite a discrepancy for a weather forecaster. An important breakthrough was made in 2002 by Falkovich and his colleagues, A. Fouxon and M.G. Stepanov. They explained how mini vortices of wind turbulence could accelerate the formation of raindrops by swirling the micrometer-sized moisture particles, thereby, allowing them to form clusters at the edges of the vortices. It is here at the edges they are more likely to collide and coalesce into larger

droplets. Scientists have calculated that clusters of approximately a millimeter are the necessary size for raindrops to fall. Falkovich says *"Ultimately, I would love to give meteorologists a simple formula that says, 'If wind is of a certain magnitude, then rain will come in forty minutes' — but this is still a dream."*[3] They do know air turbulence can affect the formation of large droplets that trigger rain. Falkovich and his team hope to be able to predict rain a few days in advance and *"within just 15 to 20 minutes and miles of its fall."*[4]

Granted rainy day predictions are becoming more accurate. As more parameters are identified and included into the picture, mathematical models and sophisticated computer simulations become more and more reliable. Will weather forecasters eventually be able to put aside their charts showing cold versus warm fronts, jet streams and satellite pictures? Will rainy days eventually be predicted from an equation involving tiny whirlwinds? Or will complexity theory of weather uncover more and more parameters which affect rainfall?

---

[1]These parameters are from an equation formulated by G. Falkovich, A Fouxon, M.G. Stepanov, from *Nature* volume 419, 9/12/02, p.152. For additional parameters and their definitions go to http://cees.tamiu.edu/cees/weather/parameters.html.

[2]*How Raindrops Form* by Geoff Brumfiel, Phys. Rev. Lett. 86, 2790 (print issue of 26 March 2001)©2001 Photodisc, inc.

[3]Ibid

quipu...numbers & words?

The quipu, an intricate array of cords with knots hanging from a primary cord, was the device the Incas used to keep track of records, quipu in essence took the place of a written mathematical system. The quipus held such records as populations, crop production, taxes, etc.. But, did they do more? How is it that the vast Inca civilization of over 2500 miles of land encompassing Peru, Ecuador to Chile, connected via a communication system using messenger runners darting along its roads, paths and bridges, had no record of a written language?

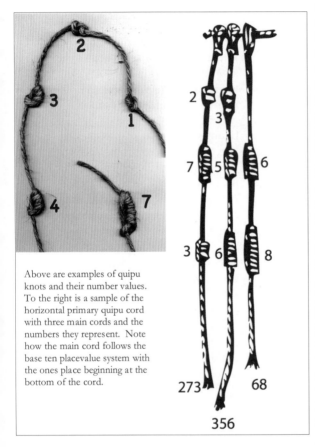

Above are examples of quipu knots and their number values. To the right is a sample of the horizontal primary quipu cord with three main cords and the numbers they represent. Note how the main cord follows the base ten placevalue system with the ones place beginning at the bottom of the cord.

In the 1990s a new theory emerged explaining that the knots of the Inca quipus did not only represent numbers, but some may have reprsented words, and thus may have also been a means of writing. Harvard anthropologist Gary Urton, an expert on the Incas and their civilization, has set out to determine if the quipus do indeed hold other non-numerical information.

affairs, and statistics for their vast empire. The knots were strategically placed on the cords so that they represented numbers by using the base ten placevalue system. The

Urton and a team of researchers— including Carrie Brezine of Harvard, cryptographers Jean-Jacques Quisquater and Vincent Castus, mathematician Vincent Blondel all of Université catholique de Louvain, computer scientists Matin and Erik Demaine of MIT—are trying to uncover whether the knots and strings of the quipus were more than a means to store numbers. Using tools of crypt-analysis similar to those used to decipher DNA's code and computer data mining techniques, the researchers are amassing a data bank at the Khipu[1] Database Project at Harvard University. It consists of information from some 750 quipus which have survived from the Inca empire and are dispersed among various museums and private collections world wide.

The above cords show both right and left twist directions and the use of colors.

The two right and left knot directions.

The different ways the main cords attach to the primary cord.

The recorded data includes specific information about the cords and their knots, such as the cords' spin or twist, how and where the cords are attached, the type of knot, the direction the knot is tied in (right or left), the knot's location on a strand, the number of types of knots and their location, the sequences of knot

numbers on the cords. Similarities between regional quipus, number sequences and their grouping are also being studied. Even the distribution of color dyes of cords is being considered, and which colors occur most frequently.

In 1923 anthropologist Leland Locke first uncovered the connection between the knots and numbers. About one-third of the extant quipus do not exactly follow the system he proposed. The quipu researchers are analyzing these "unusual" cases along with the other data collected, and are applying various computer programs to mine, search, sort, compare and tabulate this data. Some computer programs look for patterns and their frequency in hopes of discovering how these knots could also have been used to write words rather than only numbers. Even tools from *network theory* have been applied to identify if there exist relationships/connections not only between the knots of a single quipu, but among a group of quipus. At Puruchuo, an archeological site outside of Lima,Peru three quipus were found together. The researchers discovered that the respective information of these quipus seems to be linked so that the data on one may have been carried forth into another quipu from the group. With this discovery, the researchers hit upon the idea of considering a knot as a *node* of a network and the quipu itself as a *network* with *links* between networks (quipus) being the common lengths of cords. An al-

These illustrations were drawn sometime between 1583 and 1613 by D. Felipe Poma de Ayala, a Peruvian Indian. The left hand corner in the left illustration shows maize kernels used in a type of abacus counting devise referred to by Acosta in his book *Historia Natural Moral de las India*. The right scene shows an Inca checking quantities stored with a quipu accountant called a Quipucamayocs.

gorithm was devised by the Quisquater's team which worked out a way to calculate the degree of similarity that exists between nodes of two quipus(or networks) in hopes of finding knots or groups of knots that indicate a certain consistent relationship to other knots.

Urton pointed out in a 2005 interview that *"We believe that we've found a sequence of knots that was a unique signifier. We suggested it may be a place name, but we're not fixed on that particular idea. It could also be the name of the khipu keeper who made them, or the subject matter, or even a time designator."* [2]

Some anthropologists are intrigued by Urton's idea and findings, while others remain skeptical.

Early descriptions and explanations of quipus date back to the 16th century. Missionary José de Acosta,

who lived among the Inca from 1571-1586, wrote in his book *Historia Natural Moral de las Indias* published in 1596 in Spain that *"To see them use another kind of calculator, with maize kernels, is a perfect joy. In order to carry out a very difficult computation...Indians make use of their kernels. ..."*. Acosta also described how the Inca would use the quipu as a way to store events or even their biography, and later use the quipus as a reference to recall facts and sequences of events. Garcilaso de la Vega (1501-1536) wrote the following on the uses of quipus: *"...The method of making these reports was by means of knots, made of various colours, where knots of such and such colours denote that such and such crimes had been punished. Smaller threads attached to thicker cords were of different colours to signify the precise nature of the punishment that had been inflicted."* [3]

Thus it seems that colors were used to represent both abstract and concrete ideas. The same color might have stood for different ideas, just like some words derive their meaning from the context of the sentence. A white string might have represnted *sheep* or the idea of *peace,* while a red string *war.* In fact, 31 different colors have been counted on the various surviving quipus.

The Inca also attached subsidiary strings that were tied midway to a main cord rather than to the primary horizontal cord to which the main cords were attached. It seems that such subsidiary strings may have had various meanings. It could be an addition or correction to the main string, or signify a word, place or idea.

It's not farfetched to consider the knots' dual nature of words and numbers. For example, we write a number using words, 17 as seventeen, why not do the reverse and use numbers to write words? The ancient Greeks used their alphabet to write numbers. The Hebrew alphabet can also be read as numbers. *"Many passages in the Torah suggest very strongly that the scribes or authors of these ancient texts were familiar with the art of coding words according to their numerical value."[4]*

Hopefully the mystery of the quipus will soon be unraveled.

Mathematics and computer science have developed and adapted techniques to shift, sort, match data and sequences of information. One such area is *networks* in which a diagram can illustrate the connectivity between objects. A network can be something as simple as this diagram, which shows four nodes (A, B, C, D) and each with three links. For example, A is linked to be B by segment AB.

Networks can become complicated very quickly with more and more nodes connecting in a vast variety of ways. When considering a quipu a network composed of their knots (i.e. nodes) connected by cords and their various characteristics, a quipu network can be very large. The job of unraveling, comparing, and looking for connections between quipus can become huge. Various math tools can be used to cut through the data. One such tool the quipu research team is using is the *suffix tree.* The terms prefix and suffix refer to the position of data at nodes along the strings of the tree. The suffix tree can be used to trim or prune the branches (strings) of the tree to help eliminate superfluous data and find essential data and characteristics quickly.

---

[1] *Quipu* is also written with the spelling *khipu.*

[2] From *Unraveling a knotty Inca puzzle,* an interview of Gary Urton appearing in *Archaeology.org* Volume 58 Number 6, November/December 2005.

[3] G G Joseph, *The Crest of the Peacock* (London, 1991).

[4] *The Universal History of Numbers* by Georges Ifrah. Published by John Wiley & Sons, Inc. 2000, Canada. p. 239.

solutions section

page 91; Stacking squares.

page 5:
Sam Loyd voting problem
Let w = number of votes for the winner.
w+A+B +C=5,219   Therefore,
w+(w − 22)+(w − 30)+(w − 73)
    =5,219
4w = 5344
w = 1336

page 90: More to Less puzzle

page 91: Whose bill is this?
Bear in mind that the person's statement to whom it belongs must be a lie.
• If it belongs to Tom, Tom's statement is false, and the other two also are false. But one of the statements is true because one always tells the truth.
• If it belongs to Dick, his statement is a lie, which makes Tom's also false while Harry's is true. There is no contradiction assuming it's Dick's.
• If it belongs to Harry, Harry's statement has to be false. Tom's is false because it's Harry's, and Dick's is false because Tom's is false. So no true statement results if we says it's Harry's.

Therefore, the $10 belongs to Dick.

p.119
The number-check puzzle
The three groups are 715, 46 and 32890, giving:
715x46=32890.

Friar's puzzle
The upper solution has 4 even rows, 4 even columns and 8 even diagonals for a total of 16.

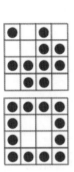

The lower solution shows 4 even rows, 4 even columns, and 10 even diagonals for a total of 18.
Note the diagonals are not restricted to the squares' diagonals.

p.119 Latin square solution

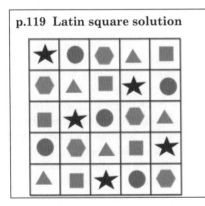

page 118: Logic problem
If any of the men had answered **no** to the first question, then there would have had to be two tan hats. Since all three answered **yes**, at least two of the men were wearing black hats. When the second round of questioning began, the first man then realized if the third man had seen a tan hat, he would have known his was black because there could not be two tan hats. So the first man concluded the third man must have seen two black hats. So he now knew his was black.

page 118: Which is which?
To determine if every black card has a square on it, we would have to turn over the black card and the card showing the circle. If the black card has a circle, then the answer is no. We need not be concerned with what the red card has on its other side, since we are dealing only with what the black card has. The card with the square can be either black or red on the oppposite side. If it is red it does not disprove that every black card has a square, and if it is black it emphasizes that every black card has a square on its opposite side.

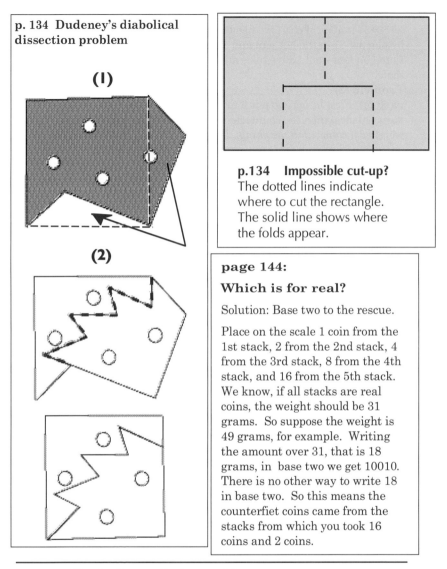

p. 134 Dudeney's diabolical dissection problem

(1)

(2)

---

p.134 Impossible cut-up?
The dotted lines indicate where to cut the rectangle. The solid line shows where the folds appear.

---

page 144:

Which is for real?

Solution: Base two to the rescue.

Place on the scale 1 coin from the 1st stack, 2 from the 2nd stack, 4 from the 3rd stack, 8 from the 4th stack, and 16 from the 5th stack. We know, if all stacks are real coins, the weight should be 31 grams. So suppose the weight is 49 grams, for example. Writing the amount over 31, that is 18 grams, in base two we get 10010. There is no other way to write 18 in base two. So this means the counterfiet coins came from the stacks from which you took 16 coins and 2 coins.

---

page 144 Claude Gaston Bachet problems
Problem (1): Since a balance scale's pans can be in just three possible states (balanced, up/down, or down/up) the first four place values of base three, namely (27, 9, 3, 1) can be used to solve this problem. For example, using weights 27oz., 9oz., 3oz., 1oz. in order to weigh something 25 oz., place 27 & 1 in left pan and 3 in the right. The additional weight you place in the right pan to balance the scale must be 25oz..
Using the balance scale and the weights 27, 9, 3, 1, every weight from 1 to 40 can be determined. Note, using all four weights gives

27+9+3+1=40oz. For 41 we need to use the next placevalue, namely 81 by placing 81 in the left pan and 1, 3, 9, & 27 on the right pan, 41 additional ounces would have to be added in the right pan to balance the scale.

Problem (2)_____

case (a) *First weighing:* put four coins on each side of the scale. If the pans balance, then the counterfeit coin is one of the remaining unweighed 4 coins. *2nd weighing:* leave 3 coins from the 1st weighing in one pan and place 3 of the unweighed coins in the other pan. If they balance, the countefeit coin is the unweighed one. *3rd weighing:* Put the unweighed coin in a pan with one of the good coins. This will tell you if its lighter of heavier.

case (b) *First weighing:* put four coins on each side of the scale. If the pans balance, then the counterfeit coin is one of the remaining 4 coins on the scale. *2nd weighing:* leave 3 coins from the 1st weighing in one pan and place 3 of the unweighed coins in the other pan. If they do not balance, the countefeit coin is of the the three you just placed in the pan. Also, the tilt of the scale tells you if the counterfeit coin is lighter or heavier. *3rd weighing:* weigh 2 of the 3 questionable coins against each other. If they balance the counterfeit is the unweighed questionable coin. If they don't balance, you know which is counterfeit from the information of light versus heavy you got from the second weighing.

case (c) *First weighing:* put four coins on each side of the scale. If the pans do not balance, then the counterfeit coin is one of the eight coins on the scale. *2nd weighing:* Now place 3 coins in each pan as follows: Split up the 4 coins in the heavier stack between the two pans and put an additional coin from the lighter pan and an unweighed coin(which we know is not counterfeit since it was not used in the 1st weighing). Now each pan has 3 coins. *There are three possibilities:*

1) If the scale balances, then one of the remaining 3 coins from the light stack that was not used in the second weighing is counterfeit, and will be lighter, since all the heavier coins were used in the 2nd weighing.. Determine which of these is counterfeit with the 3rd weighing by seeing if two of these balance. If they balance, then the unweighed coin is the lighter counterfeit coin. If they don't balance, the lighter one on the scale is the counterfeit coin.

2) If the scale doesn't balance and the pan with the noncounterfiet coin is in the heavy pan, then one of those two heavy coins is counteriet. The 3rd weighing: weigh them against one another to find the heavy counterfiet coin..

3) If the scale doesn't balance, and the pan with the noncounterfiet coin is in the light pan, then one of those two light coins is counteriet. The 3rd weighing: weigh them against one another to find the light counterfiet coin.

17-sided polygon, construction 42-43
2's properties 104-105
5th dimension, infinite warped 109
65537-gon, construction 43

A
A New Kind of Science 68
Abel Prize 137
Abel, Niels Hendrik 76, 135-137
Acosta, José de 184
Adams, Michael 89
Agrawal, Mandindra 115
algorithm 73
Alhambra 13
alpha-beta pruning 87
American Institue of Mathematics 13
Ames, Adelaide 121
Anamorph Me! 84
anamorphic art 82-84
angle of incidence 84
angle of reflection 84
anyon 111
Antikythera mechanism 126-129
Appel, Kenneth 37, 176
approval voting 34
archeological digital blueprinting 153
Archimedes 168, 167
Archimedes spiral 169
Aristotle 22
Arrow, Kenneth 34
art of Michael Burke 164-166
artificial life program 8
artificial surfparks 70
astronomical clock 63
atmosphere 147
atomic clock 63, 64
autoinducers 106, 107
Ayala, D. Felipe Poma 184

B
bacteria communication 106-107
balance scale problems 144
Barabási, Albert-László 10
Barlow's theorem 93
Bassler, Bonnie 106, 107
Bayes' theorem 27
Bayes, Thomas 27
Bayesian probability 27
bee flight 52

Beever, Julian 84
Bellard, Fabrine 115
Bergmann, Uwe 172
Bernini, Gian Lorenzo 3
Big Bang theory 112, 131
bit 111
bioinformatics 107
biomathematics 59
biometrics 47
black swan 32-33
Black, Duncan 34
Black, Kerry 70
BladeRunner 127
Blondel, Vincent 182
boids 8
Bolyai, János 81
Borda count 35
Borda election 5
Borwein, Jonathan 115
boson 130
bounded look ahead approach 87
Brahe, Tyco 53
brane 108, 109, 133
brane, closed 109
brane, open 109
Brezine, Carries 182
Brown, Greg 82
Brunelleschi dome 56-58
Brunelleschi, Filippo 57-58
buckyball 138,142,143,176
bulk 108
Bureau International des Poids et
Measures (BIPM) 62
Burke, Michael 164-166
butterfly effect 146

C
CAD(computer-assisted design) 153
candy box problem, 118
Cantor, Georg 122
carbon dioxide 147
Cardano, Girolamo 117
Case of Lies, The 25-26
casting out nines 145
Castus, Vincent 182
Cathedral of Santa Maria del Fiore 57
Cauchy, Augustin 76, 136, 137
cellular automaton 60, 68, 69
chaos theory 149

chess, history 38, 87
cholerae 106, 107
Chung, Fan 170
Cicero 129
Clay Institute 21, 26
closed strings 108, 109
colonnades 3
combinatorics 172, 175
complexity 149
computational biology 107
computer chess 87-88
Computer History Museum 154-155
computer, alpha-beta pruning 87
computer, bounded look ahead
    approach 87
computer, Deep Blue 88, 89
computer, Deep Thought 88
computer, Fool's Mate 87
computer, game trees 87, 88
computer, heuristics 88
computer, Hydra 88,89
computer, minimax algorithm 87
computer, static evaluation 87
Condorcet election 34
Condorcet, Marquis de 34
constants 96-98
continuum 23
Coordinated Universal Time (UTC) 63
Copernicus 53
cosmological constant 14
cosmology 53-54
    bulk 108
    brane 108, 109, 133
    brane, closed 109
    brane, open 109
counting numbers spiral 115
Coxeter, Harold 43
Crelle, August 136, 137
Cutler, William 170
cyclical number 124, 125

D_____
Dali, mathematics 9-11
Dali, Salvador 164
Dali, Salvador 9-11
Dali, Slave Market 9
Dali, The Crucifixition 10
Damanik, David 94

dancing and mathematics 16
DARPA 162, 163
data mining 177-179
data searches 46
Davis, Richard L. 78
decoherence 110
Deep Blue 88, 89
Deep Thought 88
Demaine, Erik 174, 182
deterministic tools 40
Diaconis, Persi 170, 172
dice 30, 31
digital blueprinting 153
digital forensics 6-7
dimensional computational geometry 8
Dimsdale, Jerry 153
dissection problems 134
distributed computing 60
document server 46
dome 56-58
Dreifus, Claudia 7
Drexel, Eric 138,139, 140, 141, 142
duality symmetry 133
Dudeney's dissection problem 134
Dürer, Albrecht 82
dynamic system 149

E_____
Earth's air-conditioning system 150
Einstein, Albert 14, 132
electromagnetism 102, 108, 133
elliptic geometry 17
emergence 60
emergence distributed computation 60,61
emergent behavior 8
Energy Innovations 95
ENIAC 155
Enigma 155
entropic system 39
Eratosthenes 114
Erdos, Paul 18, 20
Escher, M.C. 79, 156-158
eternal chaotic inflationary theory 112,113
Euclid 5th postulate 54, 81,90
Euclid's Parallel postulate 90
Euler, Leonhard 18, 92, 105
Euler's polyhedra theorem 92,105,176

F

Falkovich, Gregory 181
feddan 75
Feigenbaum constant 96, 166, 167
Feigenbaum, Mitchell 96, 169
Fermat's Last Theorem 91, 105
fermions 130
Feynman, Richard 138
Fibonacci 97, 123
Fibonacci Kecak 124
Fibonacci sequence 97, 98, 114
Fibonacci waltz 123
Field Medals 21
Finagle's Law 71
five platonic solids 54
five-fold symmetry 93-94
Fool's Mate 87
Foresight Institute 138, 140, 141
formal concept analysis 178
four color map problem twist 37
Four Color Map Theorem 176
Fourier, Jean Baptiste 51, 147
Fouxon, A. 181
fractals 65-67, 73, 112
fractals, geometric 65, 73
fractals, random 65
fractals, statistical self-similarity 65
Freeth, Tony 129
Frege, Gottlob
French Academy of Science 136, 137
Friedmann, Alexander 14
Froberg, Ted 123
Fry, John 13
fullerene 142
Fuse pot 176
Fuse, Tomoko 176

G

Galois, Evariste 76
game trees 87, 88
Gaspar, Claude 144
Gauss, Karl 42-43, 80
Gell-Mann, Murray 133
General Relativity(GR) 131, 132
geoengineering 45
Georg Pick 171
Georgia Dome 57
Girard, Albert 117

global warming 146-149
global warming effects 151-152
Goldbach conjecture 100, 104
Goldbach, Christian 100
Golden Bell Tower 124
golden ratio 96, 97
Gore, Al 149
Graham, Ron 170
Grand Canyon 41
Grand Egyptian Museum(GEM) 72-74
graph 4
graph theory 175
graphs of global warming 150-152
graviton particle 108, 109
gravity 102, 108, 109, 133
Great Pyramid of Giza 75
Green, Michael 130
greenhouse effect 147, 148, 149
Gregorian calendar 64
Griffith, Jack 110
Guenneau, Sébastien 162, 163
Gula, Sharbat 47

H

Haken, Wolfgan 37, 176
hat problem 118
Heiberg, Johan Ludwig 168
Heneghan-Peng Architects 72, 75
Heppner, Frank H. 8
Hermes, Johann Gustav 43
Hermesdorf, Kathleen 16
heuristics 88
hexahedron 100
hexa-hexa-flexagon 176
Hilbert, David 90
Hipparchus 129
Hizume, Akio 124
Homes, Susan 170
Howard, Vyvyan 142
hub 4
Hubble, Edwin 121
Hudson, Hud 34
Hull, Thomas 176
hurricane Katrina 146
Hwang, Woo-Suk 7
Hydra 88, 89
hypercube 11

I

index server 46
Indian Institute of Technology 115
infinite sequences 122
infinite series 122
infinity 122
International Earth Rotation Service 163
International Panel for Climate Change 150
internet searches 177-179
invisibility cloak 162-163
iteration 73

J

Jacobi, Carl 137
Jacquard loom 155
Joannopoulos, John 86
Jones-Smith, Katherine 66-67
juggling 28-29

K

Kacyra, Ben 153
Kaekawa's Theorem 174
Kaluza, Theordr 132
Karalis, Aristeidis 86
Karkalos, C. 126
Kasparov, Garry 89
Kawasaki's Theorem 174
Kayal, Neeraj 115
Kepler, Johannes 53, 54
knots, mathematical 110-111
Königsberg bridge problem 18-19, 91,105
Krasner Foundation 66

L

Landau, Ellen 65,67
Lang, Robert 175
Lange, Titia de 110
lattice polygon 171
lattice theory 178
lattice, mathematical 178
Lawrence Berkeley Nat'l Laboratory 15
leap second 62-63
Legendre, Andrien-Marie 136
Leibniz, Gottfried 122
Lemaitre, Georges 14
Lenstra, Hendrik W. 157, 158
Levine, Don 93
Lewis, David 70

Leyser, Keylan
Liber Abaci 123
Linde, Andrei 112, 113
Lindeman, Carl 44
line symmetry 131
Listing, Benedict 80
Lobaschesky, Nikolai 81
Locke, Leland 183
Loebner Prize 101
logic problems 119
Lucas, Edward 123
Lutaslowski, Wincenty 51
LuxS 106

M

Maekawa, Jun 176
Maekawa's Theorem 176
magic & math 12
Magnan, Antoine 52
Mahanthapp, Rudresh 124, 125
Mailgram, Stanley 4
Mandelbrot set 65, 74, 113
mapping 83
Mars Climate Orbiter 121
Martin, James 65
Mathematica 68, 69
mathematical goofs, famous 120-121
mathematical knots 110-111
mathematical lattice 178
mathematical origami 173-176
Mathematical Sciences Research Institute 99
mathematics and
    Escher's work 159
    finance and music 85
    golf ball 55
    music 124-125
    music and finance 85
    pervasiveness 160, 161
    plants 59-61
    surfing 70
    unacceptable concepts of 117
    weather 180-181
Mathur, Harsh 66-67
Matter, Alex 65-67
Maxwell, James 132, 133
McCurry, Steve 47
Merlin, Roberto 163
Messinger, Susanna 61

metamaterial 162, 163
Metonic cycle 126, 127
Millennium Dome 57
minimax algorithm 87
Miura, Koryo 175
Möbius strip 78-80
Möbius, Augustus Ferdinand 78, 80
Molecular Foundry 15
Mott, Keith 61
M-Theory 130-133
Muira Ori folding pattern 175
multiverse 112-113
Murphy's Law 71
Musheno, Thomas 47

N
nano golf balls 143
nanocatastrophes 139
nano-fabric guard 143
nano-facial cream 143
nanofactories, desktop 141
nano-glaze 143
nano-hazards 140-141
nano-rotor 142
nano-something 141, 143
nanotechnology 138-139
nanotechnology 138-143
nanotechnology desktop factory140,141
nanotechnology space bridge 103
nanotubes ropes 139
nano-water repellents 143
National Nanotech Initiative(NNI) 140
network theory 183, 185
networks 19-20
network,
  complete 19
  dynamic 19
  hub 4
  nodes 18
   regular 19
  static 19
  suffix tree 185
Netz, Revial 170
Neumann, John von 40
neutrinos 167
New Hampshire Primary 2008 27
Nicomedes' conchoid 95
nine-fold symmetry 94

nodes 18
noise 39
non-Euclidean geometries 54, 81
non-linear complex dynamic system 33
non-linear complex systems 149

O
Obadya, Philip 71
odd perfect prime problem 100
open strings 108, 109
optimization mathematics 55, 59
optimization problem 175
origami,
  buckyball 176
  flat folds 174
  mathematical 173-176
  mountain crease 174
  rigid 174, 176
  telescope 175, 176
  traditional rules 175
  valley crease 174
  vertices 174
O'Shaughnessy, Perri 25

P
palimpsest, Archimedes 168-169
paradox,
  Achilles and the tortoise 22, 24
  Arrow 23
  bouncing ball 22
  Condorcet 34
  Dichotomy 22, 24
  Stadium or Stade 22, 24
  Thomas Lamp 23
  Zeno 22-23
Parmenides 22
particles 133
Pascal triangle 104
patterns 114
Peak, David, 61
pendentives 57
Perelman, Grigory 21
perfect cuboid problem 100
permafrost 152
perspective 82
perspectoscope 82
Phidias 97
pi 44, 96, 115

Pick's Theorem 171
Pisa, Leonardo da 97, 123
pixels 6-7, 69
Plouffle, Simon 115
plurality election 5, 34
Poincaré conjecture 21
Poincaré, Henri 21, 54
point symmetry 131
polling 27
Pollock, Jackson 65-67
Pope Gregory XIII 64
Popper, Karl 32,33
Poseidonios 129
positive feedback loop 150-152
posterior probability 27
Pozzo, Andrea 116
Price, Derek de Solla 126, 127, 129
prime numbers 114, 115
Print Gallery 156-158
problem/puzzles,
    antiquity construction problems 90
    balance scale 144
    Dudeney's dissection problem 134
    Friar's puzzle 119
    Goldbach's conjecture 100
    hat problem 118
    impossible cut-up? 134
    inheritance problem 77
    Latin square puzzle 119
    More to less puzzle 90
    number check problem 119
    odd perfect prime problem 100
    Perfect Cuboid Problem 100
    Stacking the squares 91
    twin prime conjecture 100
    Which has which? 118
    Whose bill is it?
projective geometry 83
pseudo-random numbers 39,40
Ptolemy 53

Q
Qgoo 84
quantum computer 110, 111
Quantum Field Theory(QFT) 131,132
quark 133
quasicrystals 93-94
quasi-periodic music 123-125
quibit 110, 111

quintic equations 136
quipu 182-185
Quisquater, Jean-Jacques 182, 183
quorum sensing 106-107

R
Randall, Lisa 108-109
Randall-Sundrum modified model 109
Randall-Sundrum theory 108, 109
random numbers 30-31, 39-40
random sequence 31
randomness 41
Rembrandt 50-51
Rényi, Afréd 18, 20
Reynolds, Crag 8
Rick, John 153
Riemann hypothesis 25-26
Riemann, Georg 17, 25-26, 81
ring tones 68-69
Rockmore, Daniel 51
Roman arch 56, 58
rotational symmetry 93
Rukskin, Serge 21
run-off election 5, 34
Russell, Bertrand 120

S
Saari, Donald 35, 36
Saccheri, Girolamo 81
Saccheri's quadrilateral 81
Saint Peter's Square 2-3
Saint-Lague, Andre 52
Sam Loyd voting problem 5
Sant'Ignazio Church ceiling 116
Saros cycle 128, 129
Saxena, Nitin 115
Schrödinger, Edwin 94
Second International Congress of
    Mathematics 90
self-similarity 65,66, 74
sequential run-off election 34
series 122
seven-fold symmetry 94
Shanks, William 120
Shannon, Claude Elwood 28
Shannon's Juggling Theorem 29
Shapely, Harlow 121
Shechtmann, Daniel 93-94
Shuch, Erika 16

Sierpinski triangle 66, 72-74
Sierpinski, Waclaw 73
siteswaps 28-29
Skywalk, AZ 17
Smalley, Richard 139, 140, 142
social network 4
Sod's Law 71
Soijacic, Marin 86
Sonify! 85
space bridge 103
Standard Model 108
static evaluation 87
Steinherdt, Paul 93
Stephanov, M.G. 181
stereo radiography 127
Stomachion 169, 170-172
stomata 59
strings 133
Strogatz, Steven 18, 20
strong(nuclear) force 102, 108, 133
stylometry 50
suffix tree 185
Sundrum, Raman 108-109
Sunflower unit 95
supercomputing 175
superlenses 163
supersymmetry 131
surfparks 70

T
Taleb, Nicholas 32-33
Tanda, Satoshi 80
Taylor, Richard 65-67
Taylor, William 55
Temps Atomique International (TAI) 62
tessellation 93
Theory of Everything (TOE) 130-133
Thomas, Jim 143
Tipett, L.H. 30
Tombaugh, Clyde 121
topological invariants 79
topological quantum computer 110
topology 19, 83
transcendental numbers 44
translating 83
TreeMaker 175
triangulation 62
trompe l'oeil 116

Turing test 101
Turing, Alan 101
Turing's imitation game 101
twin prime conjecture 100
two color origami theorem 174

U
Uchizono, Donna 16
Ulam Stanislaw 31, 115
Urton, Gary 182,183, 184, 185

V
Vega, Garcilaso de la 184
Vibrio bacteria 106
Vibrio fisheri 106, 107
Vinci, Leonardo da 48-49, 82
Vinci, Leonardo da museum 48-49
Viswanath constant 98
Viswanath, Divakar 97, 98
voting 5
voting paradoxes 34-35

W
Walba, David M. 79
Walser, Rodger 162
Watts, Duncan 18, 20
wave theory 70
wavelet 51
wavelet transform 51
weak(nuclear) force 102, 108, 133
weather and mathematics 180-181
web server 46
West, Jevin 61
Wilson, Ngel 170
window 51
witricity 86
Witten, Edward 130
Wolfram, Stephen 68-69
Wolman, David 45
wormhole 17

X
x-ray tomography 127
X-Tek 127

Z
Zeno 22-23
zeta function 25

— About the Author —

Theoni Pappas is passionate about mathematics. A native Californian, Pappas received her B.A. from the University of California at Berkeley in 1966 and her M.A. from Stanford University in 1967. She taught high school and college mathematics for nearly two decades, then turned to writing a remarkable series of innovative books which reflect her commitment to demystifying mathematics and making the subject more approachable. Through her pithy, non-threatening and easily comprehensible style, she breaks down mathematical prejudices and barriers to help one realize that mathematics is a dynamic world of fascinating ideas that can be easily accessible to the layperson.

Her over 18 books and calendars—including *The Joy of Mathematics, Fractals & Googols & Other Mathematical Tales, Math Talk, Math Stuff*—appeal to both young and adult audiences, and intrigue the "I hate math people" as well as the math enthusiasts. Three of her books have been *Book-of-the Month Club™* selections, and her *Joy of Mathematics* was selected as a *Pick of the Paperbacks*. Her books have been translated into Japanese, Finnish, French, Slovakian, Czech, Korean, Turkish, Russian, Thai, simplified and traditional Chinese, Portuguese, Italian, and Spanish.

In 2000 Pappas received *the Excellence in Achievement Award* from the University of California Alumni Association. In addition to mathematics Pappas enjoys the outdoors, especially the seashore where she has a home. There she bicycles, kayaks, hikes and swims. Her other interests include watercolor painting, photography, music, cooking and gardening.

Other Mathematics Titles by Theoni Pappas

MATH STUFF
$12.95 • 160 pages •illustrated • ISBN:1-884550-26-6

MATHEMATICAL FOOTPRINTS
$10.95 • 160 pages •illustrated • ISBN:1-884550-21-5

MATHEMATICAL SCANDALS
$10.95 • 160 pages •illustrated • ISBN:1-884550-10-X

THE MAGIC OF MATHEMATICS
$12.95 • 336 pages •illustrated • ISBN:0-933174-99-3

FRACTALS, GOOGOLS, and Other
Mathematical Tales?
$10.95 • 64 pages • for all ages • illustrated • ISBN:0-933174-89-6

MATH FOR KIDS & Other People Too!
$10.95 • 140 pages • illustrated • ISBN:1-884550-13-4

ADVENTURES OF PENROSE
The Mathematical Cat
$10.95 • 128 pages • illustrated • ISBN:1-884550-14-2

THE FURTHER ADVENTURES OF PENROSE
The Mathematical Cat
$10.95 • 128 pages • illustrated • ISBN:1-884550-32-0

THE JOY OF MATHEMATICS
$10.95 • 256 pages • illustrated • ISBN:0-933174-65-9

MORE JOY OF MATHEMATICS
$10.95 • 306 pages • illustrated • ISBN:0-933174-73-X
cross indexed with *The Joy of Mathematics*

MUSIC OF REASON
Experience The Beauty Of Mathematics Through Quotations
$9.95 • 144 pages • illustrated • ISBN:1-884550-04-5

MATHEMATICS APPRECIATION
$10.95 • 156 pages • illustrated • ISBN:0-933174-28-4

MATH TALK
mathematical ideas in poems for two voices
$8.95 • 72 pages • illustrated • ISBN:0-933174-74-8

THE MATHEMATICS CALENDAR
$10.95 • 32 pages • written annually • illustrated • ISBN:1-884550-

MATH-A-DAY
$12.95 • 256 pages • illustrated • ISBN:1-884550-20-7

WEBSITE:
http://www.wideworldpublishing.com